T0075035

# Culture, Creativity and Economy

This book nuances our understanding of the contemporary creative economy by engaging with a set of three key tensions which emerged over the course of eight European Colloquiums on Culture, Creativity and Economy (CCE): 1) the tension between individual and collaborative creative practices, 2) the tension between tradition and innovation, and 3) the tension between isolated and interconnected spaces of creativity.

Rather than focusing on specific processes, such as production, industries or locations, the tensions acknowledge and engage with the messy and restless nature of the creative economy. Individual chapters offer insights into poorly understood practices, locations and contexts such as coworking spaces in Berlin and rural Spain, creative businesses in Leicester and the role and importance of cultural intermediaries in creative economies within Africa. Others examine the nature of trans-local cultural flows, the evolving "field" of fashion, and the implications of social media and crowdfunding platforms.

This book will be of interest to students, scholars and professionals researching the creative economy, as well as specific cultural and creative industries, across the humanities and social sciences.

**Brian J. Hracs** is Associate Professor of Human Geography at the University of Southampton, UK. He is interested in how digital technologies and global competition are reshaping the creative economy.

**Taylor Brydges** is a postdoctoral researcher at Stockholm University, Sweden and the Institute for Sustainable Futures at the University of Technology Sydney. Her research interests include economic competitiveness and entrepreneurship in the cultural and creative industries, and sustainability and the circular economy.

**Tina Haisch** is Professor for Innovation and Space at University of Applied Sciences and Arts Northwestern Switzerland. Her research focuses on processes of valuation in creative and cultural industries, how cities and regions transform through arts and culture and which role societal values play.

**Atle Hauge** is Professor in Service Innovation at Inland Norway University of Applied Sciences. From 2015 to 2019 he was the leader of Knowledge Works—the Norwegian national Centre for Cultural Industries. He has worked on several projects on the cultural industries, and his PhD thesis was on the Swedish fashion industry. Other research interests are service innovation, digitisation and regional development.

**Johan Jansson** is Associate Professor at the Department of Social and Economic Geography, Uppsala University. Jansson's research concerns the spatial organisation of (economic) activities, spatially and socially embedded processes (e.g. knowledge, creativity, values) and how technology alter distance/proximity dynamics.

**Jenny Sjöholm** is Lecturer at Linköping University in Sweden at the Department for Technology and Social Change. Her research is found in the area of geohumanities and concerns the geographies, politics and practices of contemporary art, cultural work and digitisation.

## The Dynamics of Economic Space

This series aims to play a leading international role in the development, promulgation and dissemination of new ideas in economic geography. It has as its goal the development of a strong analytical perspective on the processes, problems and policies associated with the dynamics of local and regional economies as they are incorporated into the globalizing world economy. In recognition of the increasing complexity of the world economy, the Commission's interests include: industrial production; business, professional and financial services, and the broader service economy including e-business; corporations, corporate power, enterprise and entrepreneurship; the changing world of work and intensifying economic interconnectedness.

**Agritourism, Wine Tourism, and Craft Beer Tourism**
Local Responses to Peripherality Through Tourism Niches
*Edited by Maria Giulia Pezzi, Alessandra Faggian, and Neil Reid*

**Rural-Urban Linkages for Sustainable Development**
*Edited by Armin Kratzer and Jutta Kister*

**Beyond Free Market**
Social Inclusion and Globalization
*Edited by Fayyaz Baqir and Sanni Yaya*

**Culture, Creativity and Economy**
Collaborative practices, value creation and spaces of creativity
*Edited by Brian J. Hracs, Taylor Brydges, Tina Haisch, Atle Hauge, Johan Jansson and Jenny Sjöholm*

For more information about this series, please visit: www.routledge.com/

# Culture, Creativity and Economy

## Collaborative Practices, Value Creation and Spaces of Creativity

**Edited by Brian J. Hracs, Taylor Brydges, Tina Haisch, Atle Hauge, Johan Jansson and Jenny Sjöholm**

LONDON AND NEW YORK

First published 2022
by Routledge
2 Park Square, Milton Park, Abingdon, Oxon OX14 4RN

and by Routledge
605 Third Avenue, New York, NY 10158

*Routledge is an imprint of the Taylor & Francis Group, an informa business*



*British Library Cataloguing-in-Publication Data*
A catalogue record for this book is available from the British Library

*Library of Congress Cataloging-in-Publication Data*
A catalog record for this book has been requested

ISBN: 978-1-032-05327-1 (hbk)
ISBN: 978-1-032-05330-1 (pbk)
ISBN: 978-1-003-19706-5 (ebk)

DOI: 10.4324/9781003197065

Typeset in Times New Roman
by Apex CoVantage, LLC

**This book is dedicated to the members of the CCE Network.**

# Contents

# Figures

# Tables

# Contributors

**Vasilis Avdikos** is Assistant Professor at the Department of Economic and Regional Development, Panteion University, Athens, Greece. His research focusses on creative and cultural industries and their links with processes of economic development in cities. Currently he is working on the impacts of collaborative workspaces in the urban context, cultural policies and new valuation methodologies for the socio-economic impacts of culture.

**Taylor Brydges** is a postdoctoral researcher at Stockholm University, Sweden and the Institute for Sustainable Futures at the University of Technology Sydney. Using the fashion industry as a case, her research explores economic competitiveness, digitalisation, innovation and entrepreneurship in the cultural and creative industries. More recently, her research has a focus on sustainability and the circular economy.

**Ignasi Capdevila** is Professor at the Paris School of Business and associate researcher at MOSAIC, BETA, and at the Chair NewPIC (PSB). His research interests include localised knowledge dynamics, knowledge communities, creativity and innovation management in organisational and urban contexts. Current research includes the innovation dynamics in collaborative spaces and knowledge dynamics, creative and innovation processes taking place in cities and creative industries.

**Roberta Comunian** is Reader in Creative Economy, Department for Culture, Media and Creative Industries, King's College London. She is interested in cultural policy, cultural and creative work and creative higher education. She recently coordinated an AHRC research network on Creative Economies in Africa and is currently involved in the H2020 EU funded project DISCE: Developing inclusive and sustainable creative economies.

**Marianna d'Ovidio** is Associate Professor at the University of Milan-Bicocca. Her research interests include cultural economy, creativity, social and cultural innovation and their interactions with the urban. Publications include the analysis of CCIs in the local development of urban regions, social innovation and DIY practices. She is currently involved in the H2020 Eu founded programme CICERONE: Creative Industry, Cultural Economy pROduction NEtwork.

**Mina Dragouni** is Research Associate at the Department of Economic and Regional Development, Panteion University in Greece and Adjunct Lecturer, Cultural Economics and Management at Open University, Cyprus. Her research interests include heritage and museums management, economics and grassroots cultural synergies. Her recent work focuses on heritage "in the making" and on carnival performances as collective cultural expressions of resistance.

**Lauren England** is Baxter Fellow in Creative Economies at Duncan of Jordanstone College of Art & Design, University of Dundee. Her research interests include cultural and creative work, sustainable creative business development focusing on craft and design and emerging creative economies in Africa. She is currently researching the impact of Covid-19 on urban creative economies and creative workers.

**Anna Gavanas** is Associate Professor (docent) with a PhD in Social Anthropology. She recently published a book on Swedish dance music history (*Från Diskofeber till Rejvhysteri*) and is affiliated with the Swedish Labour Movement's Archives and Library. Gavanas' research fields include migration, policy anthropology, welfare states and working life in the EU as well as Swedish retirement migrants in Spain.

**Rachel Granger** is Professor of Urban Economies, De Montfort University. Rachel specialises in urban economies, with research interests in the economic geography of creative, smart and sharing cities, and enacting inclusive urban growth. Rachel is a lead on Leicester's Flokk Lab, Urban Innovation Lab and Leicester Citizen Sensor, as well as supporting the economic recovery strategy and innovation strategy for Leicestershire.

**Tina Haisch** is Professor in Innovation and Space at University of Applied Sciences and Arts, Northwestern Switzerland. Her research focuses on processes of valuation in creative and cultural industries and how these processes are changing existing geographies of production. She analyses how cities, neighbourhoods and regions transform through arts and culture and how societal values like tolerance impact regional economic development.

**Atle Hauge** is Professor in Service Innovation at Inland Norway University of Applied Sciences. From 2015 to 2019 he was the leader of Knowledge Works—the Norwegian National Centre for Cultural Industries. He has worked on several projects on the cultural industries, and his PhD thesis was on the Swedish fashion industry. Other research interests are service innovation, digitisation and regional development.

**Brian J. Hracs** is Associate Professor of Human Geography at the University of Southampton. He is interested in how digital technologies and global competition are reshaping markets, labour conditions and spatial dynamics within the creative economy. He is currently researching the processes and spatial dynamics of curation, the trans-local nature of cultural scenes and creative economies in Africa.

**Johan Jansson** is Associate Professor at the Department of Social and Economic Geography, Uppsala University. His research is on the spatial organisation of (economic) activities, spatially and socially embedded processes and how technology alter dynamics of distance/proximity. Empirically Johan focusses on cultural industries (music, theatre, arts), the internet industry, local milieus, urban/rural and regional development.

**Mariangela Lavanga** is Senior Assistant Professor Cultural Economics, Academic coordinator, MA Cultural Economics and Entrepreneurship, Co-founder and Academic coordinator, Minor Fashion Industry at Erasmus University Rotterdam. She addresses the role of cultural industries in sustainable urban development, focusing on fashion and design industries. She recently co-led the project RE-FRAME FASHION, supported under the Erasmus+ Strategic Partnership Programme of the European Union.

**Deborah Leslie** is a Professor in the Department of Geography and Planning at the University of Toronto. Her research topics are united by a common theme: the relationships between economy, culture and place. Her research interests are in the location and dynamics of cultural industries and their production systems, urban cultural policies, new forms and scales of urban governance, the spatial and temporal logic of commodity chains and networks and ethical issues surrounding consumption, the body and the workplace.

**Janet Merkel** is an urban sociologist and post-doctoral researcher at the Chair of Urban and Regional Economics, Institute of Urban and Regional Planning, Technical University, Berlin. She is a guest professor at the Institute of Urban Development, Kassel University. Her research interests include new forms of work organisation (coworking), creative labour, creative industries, cultural planning and urban politics.

**Judith Nyfeler** is a post-doctoral researcher and lecturer at the Institute of Sociology, University of St. Gallen, Switzerland. Her research interests include innovation, creativity and how communication and technology affect the making of novelties. Judith's current project focuses on the resurgence of craft(s) and the relation between industrial and handcraft manufacture methods.

**Norma M. Rantisi** is Professor of Geography and Planning at Concordia University. She is the co-editor of the online magazine *Progressive City: Radical Alternatives*. Collaborating with Dr Deborah Leslie at University of Toronto, she has started a new project on workforce intermediary social enterprises and the opportunities and challenges presented when operating within neoliberal governance and funding regimes. Dr Rantisi is also starting a project on women's artisanal cooperatives.

**Anders Rykkja** is Doctoral Research Fellow at Inland Norway University of Applied Sciences and working on his doctoral thesis about the usage of crowdfunding as a business model and valuation mechanism in the cultural and creative sector in the Nordic countries and Spain. He is part of the Norwegian Research Council project CROWDCUL on adoption and usage of crowdfunding among artists.

**Jenny Sjöholm** is Lecturer at Linköping University in Sweden, Department for Technology and Social Change. Her research is focused on geohumanities and the geographies, politics and practices of contemporary art, collections and cultural work. She is currently involved in two projects: one examining the construction of value in the art market and the second exploring memory work as gendered emotional labour.

**Lech Suwala** is Guest Professor of Urban and Regional Economics at Technische Universität Berlin, previously Visiting Professor in Innovation/Creativity Management, Ritsumeikan University, Osaka and was a Research Associate at the Institute of Economic Geography at Humboldt-Universität zu Berlin. Lech's research includes spatial creativity, innovation and entrepreneurship research, European and regional planning, as well as place-based policies at various levels.

# Acknowledgments

This edited collection emerged from eight European Colloquiums on Culture, Creativity and Economy (CCE) which not only served as fruitful events but the genesis of a supportive and collaborative network of creative economy researchers. The editors would like to thank all of the network members, who are listed in Table (XX), for their ideas, inspiration, invitations and collaborations which have not only been so valuable to this book but our lives and careers more broadly.

By extension, we would like to acknowledge the CCE creators—who were all based at Uppsala University at the time—for dreaming up the colloquium (Brian J. Hracs, Johan Jansson, Jenny Sjöholm and Dominic Power) and in turn, all of the CCE series organisers for all the time and effort involved in putting on these intensive and energising events (Taylor Brydges, Carol Ekinsmyth, Atle Hauge, Brian J. Hracs, Johan Jansson, Cecilia Pasquinelli, Dominic Power, Suntje Schmidt, Jenny Sjöholm and Suzanne Reimer).

As these colloquiums were shaped by the local flavours, sights and sounds of the spaces and cities where each event took place we would also like to thank the local organisers for being gracious hosts and expert "curators": Brian J. Hracs, Johan Jansson, Jenny Sjöholm and Dominic Power (Uppsala, 2012), Oliver Ibert and Lech Suwala (Berlin, 2013), Barbara Heebels and Robert Kloosterman (Amsterdam, 2014), Francesco Capone, Niccolò Innocenti and Luciana Lazzeretti (Florence, 2015), Taylor Brydges, Atle Hauge and Anders Rykkja (Seville, 2016), Brian J. Hracs (London, 2017), Taylor Brydges, Johan Jansson, Jenny Sjöholm and Dominic Power (Stockholm, 2018) and Tina Haisch and Judith Nyfeler (Basel, 2019).

Putting together an edited collection involves many steps and many moving parts. We would like to thank all of the contributors to the book for their efforts in generating novel ideas, meeting deadlines and responding to editorial requests. The book also benefited from valuable comments from Charlotte Campbell, Lauren England and Timo Koren, who we turned to for feedback throughout the process. We are also grateful to our project manager Andrea Tadros for all of her effort in driving the process forward and always keeping us on our toes. We as editors (Atle, Jenny, Johan, Taylor and Tina) would also like to direct a special thank you to our co-editor Brian J. Hracs for his consistent organisational work and effort on all of the CCE events and this book.

Finally, we are grateful for the financial support provided by The Swedish Research Council (Vetenskapsrådet—Rambidrag för kulturforskning), under the project title "Culture, Creativity and Economy", and Knowledge Works, which is a program financed by the Norwegian Ministry of Culture to develop knowledge about the cultural and creative industries (www.kunnskapsverket.org).

*Table XX* Members of the CCE Network

| | | | |
|---|---|---|---|
| Dan Ashton | Melanie Fasche | Benjamin Klement | Cecilia Pasquinelli |
| Vasilis Avdikos | Nicole Foster | Leandro Valiati | Dominic Power |
| Alison Bain | Alison Gerber | Montserrat Pareja-Eastaway | Andy Pratt |
| Mark Banks | Gernot Grabher | Stefania Galli | Rhiannon Pugh |
| Marco Bettiol | Rachel Granger | Robert Kloosterman | Norma M. Rantisi |
| Jeff Boggs | Tina Haisch | Bastian Lange | Susanne Reimer |
| Thomas Boren | Atle Hauge | Mariangela Lavanga | Josephine V. Rekers |
| Amanda Brandellero | Harriet Hawkins | Luciana Lazzeretti | Lizzie Richardson |
| Verena Brinks | Barbara Heebels | Deborah Leslie | Silvia Rita Sedita |
| Julie Brown | Ilse Helbrecht | Andrew Leyshon | Anders Rykkja |
| Taylor Brydges | David Hesmondhalgh | Peter Lindner | Suntje Schmidt |
| Hans-Joachim Bürkner | Brian J. Hracs | Marie Mahon | Susanne Schulz |
| Charlotte Campbell | Phil Hubbard | Anders Malmberg | Jenny Sjöholm |
| Ignasi Capdevila | Michael Hutter | Angela McRobbie | Peter Sunley |
| Francesco Capone | Oliver Ibert | Max-Peter Menzel | Lech Suwala |
| Patrizia Casadei | Niccolo Innocenti | Janet Merkel | Paul Sweetman |
| Karenjit Clare | Doreen Jakob | Steve Millington | Calvin Taylor |
| Nick Clifton | Al James | Felix Müller | Joachim Thiel |
| Roberta Comunian | Johan Jansson | Sean Nixon | Nicola Thomas |
| Paz Concha | Høgni Kalsø Hansen | Jakob Nobuoka | Alberto Vanolo |
| Louise Crewe | Hang Kei Ho | Massimiliano Nuccio | Tarek Virani |
| Marianna D'Ovidio | Laura Kkula-Wenz | Judith Nyfeler | Jacqueline Wallace |
| Carol Ekinsmyth | Kim-Marie Spence | Kate Oakley | Jon Ward |
| Lauren England | David Oamen | Stefania Oliva | Saskia Warren |
| | | | Anders Waxell |

# 1 Introduction

## Exploring tensions in the creative economy

*Brian J. Hracs, Taylor Brydges, Tina Haisch, Atle Hauge, Johan Jansson and Jenny Sjöholm*

The myriad links between culture, creativity and the economy are key elements of modern life and central topics of intellectual discussion (Caves, 2002; Hesmondhalgh, 2002; Power and Scott, 2004). Over the past two decades, policy makers and academics around the world have become deeply interested in a range of interconnections between cultural and economic processes: including culturally-driven economic development in the form of 'creative' cities and cultural quarters (Kloosterman, 2013; Mould and Comunian, 2014; d'Ovidio, 2016), the dynamics of creative labour (Hesmondhalgh and Baker, 2010; Bain and McLean, 2013; Pasquinelli and Sjöholm, 2015; Reimer, 2016; McRobbie, Strutt and Bandinelli, 2019), the evolution of specific cultural and creative industries such as art, music, fashion and craft (Power and Scott, 2004; Hesmondhalgh, 2002; Leyshon, 2014; Crewe, 2017; Jakob and Thomas, 2017), competition and value creation within markets (Hracs, Jakob and Hauge, 2013; Bürkner and Lange, 2017; Ibert et al., 2019) and the role of space and specificity across different locations, scales, industries and time periods (Cooke and Lazzaretti, 2007; Edensor et al., 2009 Gibson, 2011; Luckman, 2012; Flew, 2013; Gill, Pratt and Virani, 2019; Schmidt, 2019). Yet, the restless dynamism of the contemporary creative economy requires ongoing empirical study, theorisation and critical reflection (Hawkins, 2016; Banks, 2018; Mould, 2018; Oakley and Ward, 2018).

This edited volume aims to contribute to and nuance existing studies, debates and knowledge. However, rather than focusing on a specific process such as production, industry, location or scale from a disciplinary perceptive this book applies a different approach. We seek to acknowledge and engage with the messy and often contradictory nature of the cultural and creative economy. For example, creative labour is said to be both precarious and rewarding (McRobbie, 2015; Brydges and Hracs, 2019a). Furthermore, while cultural and creative industries are said to cluster in global hotspots, a growing literature also highlights the presence and uniqueness of cultural

DOI: 10.4324/9781003197065-1

and creative activities in peripheral and rural settings and across the Global South (Comunian et al., 2021). At the same time, digital transitions and the forces of globalisation and competition continue to create, destroy and restructure the markets and conditions under which cultural creation, production, promotion, intermediation, dissemination and consumption are undertaken and experienced (Hracs, 2015).

Based on these observations, we have chosen to organise this book around a set of three key tensions; 1) the tension between individual and collaborative creative practices, 2) the tension between tradition and innovation, and 3) the tension between isolated and interconnected spaces of creativity. These tensions are at the heart of the connection where culture and creativity meet the contemporary economy and where restless dynamism unfolds. The tensions provide frustration and friction, as well as energy and inspiration. Simultaneously, these tensions are ingenious and vicious. As such, they are similar to the process Schumpeter famously coined "creative destruction" (Schumpeter, 1959 [1934]). The chapters in this book deal with processes, observations and phenomena that can be located in one or more fields of these tensions.

## Collaborative practices: the story of the CCE network

The foundations of this book were laid at the European Colloquium on Culture, Creativity and Economy (CCE). Since the first meeting of creative economy researchers in Uppsala (2012) and through subsequent annual events in Berlin, Amsterdam, Florence, Seville, London, Stockholm and Basel, CCE has become a network of over 100 scholars from 19 countries who work in a range of disciplines including geography, sociology, economics and cultural studies. This book presents a range of chapters from community members who have been working collaboratively—by sharing, extending and co-producing ideas, research and publications—to develop a deeper understanding of the contemporary creative economy for nearly a decade.

The goals of the CCE meetings were manifold: firstly, we aimed to bring together a small number of scholars, between 20 and 35, from different disciplines, universities and levels of experience to develop new approaches to the creative economy. Secondly, our goal was to build networks which encouraged knowledge sharing and collaboration. Thirdly, we wanted the events to be informal, open and supportive atmospheres where trust could be built and where people could feel safe to share ideas and work in progress rather than, as is usual at conferences, finished products (see also Pratt, 2010). The events featured a range of 'blocks' including themed panels, small discussion groups, one-on-one peer feedback sessions and informal walking tours. Together, these elements produced positive environments and experiences in line with Dorling's (2019) call for a "new rigour" based on kindness. Fourthly, we strived for diversity and inclusiveness. While the majority of our participants came from across Europe they were not necessarily 'European', and we recruited widely and were also able to invite and include scholars from Africa, Asia, the Caribbean, North America and South America. Furthermore, we aimed for and achieved equal representation of men and women in the network, and a balance of scholars at different career stages coming from a range of disciplines at each event. Finally, we aimed to merge the community of practice and the community of science. To marry theoretical debates with the 'real' world—and to try and "dismantle the ivory tower" (Klein et al., 2011)—we invited to our meetings, artists, curators and other practitioners working in the cultural and creative industries.

## Three key tensions within the creative economy

Over the course of our eight CCE meetings, while there were some processes, concepts or challenges that would be introduced as themed panel topics once or twice, others seemed to re-emerge year after year in a variety of sessions and interactions. By examining and re-examining some of these ideas—such as coworking spaces or curation—from different perspectives, over time our collective understanding deepened and became more nuanced. This iterative process also exposed and encouraged community engagement with a number of the previously mentioned tensions, which not only spurred lively debates during the events but also many other academic works. For this book, we invited network members to submit chapters which address one or more of the three key tensions outlined below.

### *The tension between individual and collaborative creative practices*

In exploring the variety of ways in which work in cultural and creative fields is organised, performed and experienced, we find a diverse range of individual and collaborative practices. While existing literature points to creative teams, project ecologies and networks of actors, including intermediaries, involved with creative production (Becker, 1982; Grabher, 2002), creative endeavours are also associated with lone geniuses, including artists, and creative entrepreneurs following 'do it yourself' (DIY) models (Sjöholm and Pasquinelli, 2014; Hracs, 2015). Here, we are interested in how entrepreneurs working in precarious, individualised and under-resourced areas of the economy make space for their work and how they collaborate with other creatives and intermediaries to cope with precarity. Indeed, in a marketplace typically characterised by competition, insecurity, alienation, self-realisation, emotional and affective labour, self-exploitation and uncertainty, creative workers develop strategies and practices based on their individual characteristics, motivations and circumstances (Hesmondhalgh and Baker, 2010; Brydges and Hracs, 2019a; Banks, 2020). Analysing such practices and the role-specific spaces play helps us understand how knowledge creation and creativity are affected by relations between individuals, communities and organisations.

Several chapters in this book consider the practices and spatial dynamics associated with coworking and the myriad ways in which traditional and emerging cultural intermediaries, from gatekeepers to curators, can enable, support or constrain creativity and creative businesses in these spaces. In Chapter 5, for example, Merkel and Suwala look at technology parks and coworking spaces in Berlin to investigate how interactive atmospheres and communities can be fostered and how such environments contribute to the emergence of new ideas and different forms of working together. Complementing these finding, in Chapter 8 Capdevila focusses on how coworking spaces and practices found in urban centres, such as Barcelona, can be translated for and transported to rural settings.

With respect to the vital, yet often hidden, role played by cultural intermediaries, Comunian et al. (Chapter 9) demonstrate that many intermediaries continue to perform traditional roles such as 'connecting producers and consumers' and 'taste-making'. However, in the wake of global competition, advances in digital technologies and shifting policy agendas they also highlight how the range of intermediary actors and functions in Africa have expanded to include providing space, finance, training and business advice which are central to enabling and sustaining creative endeavours. In Chapter 10 Jansson and Gavanas focus on the specific intermediary function of curation to help understand the evolution of electronic dance music in Stockholm. By comparing two contrasting eras, they analyse the role that intermediary pioneers, spaces and processes of resistance play in sorting, filtering and contextualising new (sub)cultural expressions. Finally, by looking at the role that crowdfunding and digital spaces can play in producing symbolic value, Rykkja and Hauge (Chapter 4) identify and explore collective dimensions within cultural and creative industries such as fashion while emphasising the growing centrality of the consumer.

### *The tension between tradition and innovation*

Innovation is central to the competitiveness of cultural, goods, services and experiences (Pine and Gilmore, 2011). Indeed, innovation is constantly disrupting the value chains and processes of design, production, promotion, distribution, intermediation and consumption. Despite what appears to be an era of rapid and unprecedented growth and change, many within cultural and creative industries such as music, fashion and art cling to and celebrate roots steeped in local traditions, craft and heritage. For some industries and actors, a competitive edge lies in the decision *not* to change with the times, but instead to rely on storied reputations of quality that are deeply imbued in products and place and are difficult to replicate or upend. Thus, some producers, operating at global and local scales, generate value and distinction by invoking traditional values, techniques and spatial entanglements that cannot be scaled up, digitised or replicated (Jansson and Waxell, 2011; Crewe, 2013). This is where we see a tension between the speed of scalable production, often spurred by automation, and the deliberate slowness of traditional craftsmanship and the production of truly unique pieces.

Several chapters in this book engage with this tension by examining how values can be generated, assessed and communicated. In Chapter 6, for example Nyfeler illustrates the tension between preservation and renewal through a study of two Swiss fashion firms with different approaches to creativity, design and production. Treating the creative outcomes of fashion as 'stylistic innovations' Nyfeler asserts that variations of cultural meaning and aesthetic alteration are rarely developed from scratch but rather evolve from existing forms. In Chapter 4 Rykkja and Hauge also illustrate how products with long-standing traditions and hand-made production, such as leather handbags and watches, can thrive alongside digital practices in the contemporary marketplace. Indeed, they demonstrate how new developments, such as crowdfunding and consumer co-creation support rather than supplant traditional practices and forms of symbolic value.

Broadening the scope to reflect on a range of developments in the global fashion industry, Brydges et al. (Chapter 2) consider the ways in which digital technologies are disrupting the roles, positions and influence of actors, such as traditional fashion magazines and new social media influencers across fashion's fields. They also examine threats to established institutions such as Fashion Weeks and bricks and mortar shops from new innovations and actors disrupting retail, marketing and distribution. Avdikos and Dragouni (Chapter 7) study similar mechanisms but focus on the heritage and museum sector. They assert the need to move beyond purely economic valuations of these activities and introduce a novel conceptual framework of cultural heritage value that considers economic impacts as well as the effects on the social and cultural fabric of surrounding places and user communities. They are critical and question the all-encompassing economic logic often underpinning policies and strategies concerning cultural artefacts, practices, institutions, places and experiences within the creative economy.

### *The tension between isolated and interconnected spaces of creativity*

Traditionally, creative workers and cultural and creative industries have concentrated in a handful of global cities, such as art in New York, music in London and fashion in Paris or Milan (Power and Scott, 2004). These global cities are said to dominate other national or regional markets and have long acted as both local and global talent magnets, attracting ambitious creatives from around the world. Recently, however, there has been increased recognition of the crucial role of peripheral and vernacular spaces of creativity and creative work such as in second tier cities and rural and/or remote spaces (Edensor et al., 2009; Gibson, 2011). New and evolving digital technologies are increasingly contributing to greater mobility and enable individuals and groups to work in 'third places' or hubs which may be located far away from urban centres (Gill, Pratt and Virani, 2019; Merkel, 2019; Schmidt, 2019).

Yet, technologies and virtual channels of communication also serve to connect and integrate individuals operating in large, small, suburban, rural and remote locations into global markets and industry-specific ecosystems (Brydges and Hracs, 2019b). These developments challenge the notion that there is one 'place to be' and break down the binaries between urban and rural, as well as established and emerging industry locations. Indeed, culturally stimulating, natural and remote landscapes are becoming increasingly attractive to creatives (Haisch, Coenen and Knall, 2017) and many find ways to live, work and create in a range of spaces using forms of temporary, mediated and virtual mobility (Brydges and Hracs, 2019b).

Several chapters in this book interrogate these developments by looking at creative practices, processes and examples at different scales and within different contexts, such as in emerging and/or poorly understood locations like Switzerland and Scandinavia, as well as in rural Spain and creative centres in Africa. In Chapter 8, for example, Capdevila shows that through a process of reinterpretation, different aspects of coworking, have become disembedded from their original urban context to be re-embedded in more peripheral environments. Relatedly, Jansson and Gavanas (Chapter 10) highlight the interconnected nature of the creative economy by using the case of electronic dance music in Stockholm to trace translocal flows of subcultural expression from global cultural nodes to the periphery.

In Chapter 3 Granger demonstrates how the trajectories and experiences of specific creative economies, such as Leicester in the UK, are connected to but also distinct from global cities which are often held up as universal examples by researchers and policy makers. Here the importance of specificity and scale is emphasised and this is a theme picked up by Comunian et al. (Chapter 9) who highlight similarities and differences between the experiences of cultural intermediaries operating in the Global North and Global South by looking at the emerging and poorly understood creative economies of Cape Town, Lagos and Nairobi. Finally, through the application of Bourdieu's concept of 'the field', Brydges et al. (Chapter 2) examine the evolving spatial hierarchies within the fashion industry and the connections and power dynamics between global and 'not-so-global' fashion centres. They also highlight the overlaps and tensions between a range of physical, virtual and temporary spaces where specific fashion-related activities occur.

## Studying the restless creative economy: future research paths

Throughout this book, the authors have approached the creative economy from different theoretical, methodological and empirical perspectives. In so doing, the chapters have engaged with and contributed to our understanding of the intersections between culture, creativity and economy as well as three key tensions within the creative economy. However, no collection of studies can be comprehensive and we acknowledge the lack of in-depth discussion around important topics such as gender and diversity. A larger book would have also been able to consider a wider range of creative spaces, contexts, industries and actors.

Moreover, as the research for this book was conducted before COVID-19 emerged, the chapters do not explicitly engage with the important implications of the pandemic for the creative economy or the responses by workers, firms, consumers and policy makers around the world. At a time when many elements within markets, societies and the lives of individuals remain highly uncertain, there is a pressing need for further research related to COVID-19. However, it is important to remember that for a variety of reasons the creative economy has always featured restless dynamism and has often encountered and experienced disruptions and transitions before other sectors (Leyshon, 2014). Therefore, in this final section we would like to propose some avenues for further research, while reiterating the need for investigation beyond the current crisis.

### *New organisation of creative labour*

The nature of creative labour has been evolving since the 1990s, brought on by the rise of new technologies, industrial restructuring, globalisation, flexibilisation and individualisation (Ekinsmyth, 1999; Hracs, 2015). Thus far, the pandemic is accelerating and intensifying these developments

(Banks, 2020). As the spaces where culture is produced, enacted, performed or exhibited shutter or close, home and remote-based work is becoming more prevalent exacerbating feelings of isolation and the tensions between private and professional spheres. As the role of physical proximity changes research is needed to investigate the structures, spatial dynamics and implications of new formulations and organisations of creative labour.

For example, what sorts of spaces are suitable for working individually and collectively? What will 'collaborative' mean in the future and what is the potential for new forms of solidarity in the workplace to mediate risk and combat precarity in the sector? Moreover, it will be important to continue to recognise the role of specificity and intersectionality by examining the ways in which individual workers, in different industries and locations, negotiate the labour market based on their own unique characteristics, contexts, aspirations, strategies and experiences (Brydges and Hracs, 2019a).

### *The rise of the new intermediaries*

Beyond pure artistic expression, the cultural and creative industries are driven by the need to shape and cater to the ever-changing tastes and aesthetic desires of individual consumers. When studying specific industries and markets there has been a tendency to focus on producers and consumers rather than actors who serve as co-producers, gatekeepers, brokers, taste-makers, curators and more recently 'social media influencers'. Although the concept of cultural intermediaries has been around for decades, it remains an umbrella term for a range of actors and activities. Indeed, the exact nature of the positions that intermediaries hold within value chains and networks and the functions they perform within the marketplace for cultural and creative products remain ambiguous (Jansson and Hracs, 2018; Haisch and Menzel, 2019).

As a result, there is a need for situated case studies that explore and attempt to differentiate the practices of specific actors, including individuals, institutions, spaces, events and platforms, operating in specific spatial contexts and industries. It would also be useful to investigate how and why cultural intermediaries interact, compete and collaborate with other intermediaries across the broader creative economy. By extension, there is a growing need to examine how and why producers work with different intermediaries and the ways in which consumers are being enrolled in intermediary functions, such as curation, through different channels including social media.

### *The digital and geographical divide in the creative economy*

New technologies, including crowdfunding, social media and content streaming platforms stimulate and enable creativity through the co-creation of products and interactions between actors in distant markets. On one hand, ongoing digitalisation enables more widespread and intensive exchanges, which may increase the creative potential for some actors. On the other hand, there is exclusion and disadvantage for those who lack access to digital infrastructure, technical equipment, education, knowledge and training. As a result, there is a growing awareness that we need to address digital divides and understand that digital spaces can be just as socially and inequitably coded as physical spaces. Moreover, while digital technologies and infrastructures tend to cluster in large urban centres, many small, rural and remote areas around the world lag behind. Thus, as this book has demonstrated geography and context matter, and will continue to matter in the future, but exactly how space, place and scale shape elements of the creative economy is a key area for further and ongoing research. For example, how can inclusivity be increased and access to knowledge and markets be enhanced and equalised? How are links made between actors and communities located in distant places? Future studies on the emergence and acceleration of creativity through building bridges of language, technology and space will contribute to a better understanding of the creative economy.

### *Towards a sustainable creative economy?*

For traditional industries and emerging innovative enterprises, creativity is and will be an important element of strategies to tackle challenges associated with environmental impacts, climate change and sustainability. Indeed, although not all cultural and creative industries produce a tangible output, there is a strong need to consider the environmental impact of processes of production, distribution and consumption. Here, the 17 UN Sustainable Development Goals (SDGs) can provide a blueprint for creative economy researchers. For example, through our research, how can we contribute to goals of gender equality, decent work and economic growth as well as sustainable cities and communities? Recognising the ability of the creative economy to create and communicate a range of non-economic values, including social and environmental sustainability, is also important (Jansson and Hracs, 2018; Brydges and Hracs, 2019a). Although art, music and fashion certainly create economic value for some, the intangible values of aesthetics, identity and ways of life are also inherent qualities in the creative economy. In addition, actors in the creative economy, including individuals, firms, brands, organisations and institutions, play significant roles in producing knowledge, information, education and content to educate, inform and disseminate knowledge on climate change, environmental impacts and solutions to current and future challenges. Thus, research is needed to map the ways in which creatives and specific industries, most notably fashion, are developing, practicing and promoting forms of sustainability.

### *The creative economy in times of crisis*

Finally, amidst the ongoing COVID-19 crisis, it is impossible not to acknowledge and investigate the myriad, and potentially long-lasting, impacts of the pandemic on different aspects of the creative economy. From the shifting conditions of work and shuttering of spaces including cinemas, galleries, theatres and retail shops to the shifts in cultural consumption and policies to support, or abandon, creativity, the cultural and creative industries have been walloped. Beyond the bottom line for firms, individuals already negotiating precarious conditions are suffering further from cancelled incomes and a general lack of targeted support. Therefore, it is vital to develop projects which explore how the crisis is affecting creativity, culture, intermediation and consumption in the short term and over time. Concomitantly, ongoing research is needed which seeks to understand the role and importance of culture and creativity at times of crisis and how the cultural and creative industries contribute not only to employment and innovation but greater personal and societal wellbeing. As Banks (2020, p. 1) argues "while culture and arts may not be vital to the preservation of life, they are proving increasingly vital to preserving the sense of life being lived".

## References

Bain, A. and McLean, H. (2013). The artistic precariat. *Cambridge Journal of Regions, Economy and Society*, 6(1), pp. 93–111.

Banks, M. (2020). The work of culture and C-19. *European Journal of Cultural Studies*, 23(4), pp. 648–654.

Banks, M. (2018). Creative economies of tomorrow? Limits to growth and the uncertain future. *Cultural Trends*, 27(5), pp. 367–380.

Becker, H. (1982 [2008]). *Art worlds*. Berkley: University of California Press.

Brydges, T. and Hracs, B.J. (2019a). What motivates millennials? How intersectionality shapes the working lives of female entrepreneurs in Canada's fashion industry. *Gender, Place & Culture*, 26(4), pp. 510–532.

Brydges, T. and Hracs, B.J. (2019b). The locational choices and interregional mobilities of creative entrepreneurs within Canada's fashion system. *Regional Studies*, 53(4), pp. 517–527.

Bürkner, H.J. and Lange, B. (2017). Sonic capital and independent urban music production: Analysing value creation and 'trial and error' in the digital age. *City, Culture and Society*, 10, pp. 33–40.

Caves, R. (2002). Creative industries: Contracts between art and commerce. Cambridge: Harvard University Press.

Comunian, R., Hracs, B.J. and England, L. (2021). *Higher education and policy for creative economies in Africa: Developing creative economies*. London: Routledge
Cooke, P. and Lazzaretti, L., eds. (2007). *Creative cities, cultural clusters and local economic development*. Cheltenham: Edward Elgar.
Crewe, L. (2013). Tailoring and tweed: Mapping the spaces of slow fashion. In: *Fashion cultures: Theories, explorations and analysis*. London: Routledge.
Crewe, L. (2017). *The geographies of fashion: Consumption, space, and value*. Bloomsbury: Bloomsbury Publishing.
Dorling, D. (2019). Kindness: A new kind of Rigour for British Geographers. *Emotion, Space & Society*, 33.
d'Ovidio, M. (2016). The creative city does not exist. Critical essays on the creative and cultural economy of cities. Milano: Ledizioni.
Edensor, T., Leslie, D., Millington, S. and Rantisi, N., eds. (2009). *Spaces of vernacular creativity: Rethinking the cultural economy*. London: Routledge.
Ekinsmyth, C. (1999). Professional workers in a risk society. *Transactions of the Institute of British Geographers*, 24(3), pp. 353–366.
Flew, T. (2013). *Global creative industries*. Cambridge: John Wiley.
Gibson, C., ed. (2011). *Creativity in peripheral places: Redefining the creative industries*. London: Routledge.
Gill, R., Pratt, A.C. and Virani, T.E., eds. (2019). *Creative hubs in question: Place, space and work in the creative economy*. Cham: Springer.
Grabher, G. (2002). The project ecology of advertising: Tasks, talents and teams. *Regional Studies*, 36(3), pp. 245–262.
Haisch, T., Coenen, F.H. and Knall, J.D. (2017). Why do entrepreneurial individuals locate in non-metropolitan regions? *International Journal of Entrepreneurship and Innovation Management*, 21(3), pp. 212–233.
Haisch, T. and Menzel, M.P. (2019). *Temporary markets in a global economy: An example of three Basel art fairs* (No. geo-disc-2019_14). Vienna: Institute for Economic Geography and GIScience, Department of Socioeconomics, Vienna University of Economics and Business.
Hawkins, H. (2016). *Creativity: Live work create*. London and New York: Routledge.
Hesmondhalgh, D. (2002). The cultural industries, 1st ed. London: Sage Publications.
Hesmondhalgh, D. and Baker, S. (2010). A very complicated version of freedom': Conditions and experiences of creative labour in three cultural industries. *Poetics*, 38(1), pp. 4–20.
Hracs, B.J. (2015). Cultural intermediaries in the digital age: The case of independent musicians and managers in Toronto. *Regional Studies*, 49(3), pp. 461–475.
Hracs, B.J., Jakob, D. and Hauge, A. (2013). Standing out in the crowd: The rise of exclusivity-based strategies to compete in the contemporary marketplace for music and fashion. *Environment and Planning A*, 45(5), pp. 1144–1161.
Ibert, O., Hess, M., Kleibert, J., Müller, F. and Power, D. (2019). Geographies of dissociation: Value creation, 'dark' places, and 'missing' links. *Dialogues in Human Geography*, 9(1), pp. 43–63.
Jakob, D. and Thomas, N.J. (2017). Firing up craft capital: The renaissance of craft and craft policy in the United Kingdom. *International Journal of Cultural Policy*, 23(4), pp. 495–511.

Jansson, J. and Hracs, B.J. (2018). Conceptualizing curation in the age of abundance: The case of recorded music. *Environment and Planning A*, 50(8), pp. 1602–1625.

Jansson, J. and Waxell, A. (2011). Quality and regional competitiveness. *Environment and Planning A*, 43(9), pp. 2237–2252.

Klein, P., Fatima, M., McEwen, L., Moser, S., Schmidt, D. and Zupan, S. (2011). Dismantling the ivory tower: Engaging geographers in university—community partnerships. *Journal of Geography in Higher Education*, 35(3), pp. 425–444.

Kloosterman, R.C. (2013). Cultural amenities- large and small, mainstream and niche—A conceptual framework for cultural planning in an age of austerity. *European Planning Studies*, 22(12), pp. 2510–2525.

Leyshon, A. (2014). *Reformatted: Code, networks, and the transformation of the music industry*. Oxford: Oxford University Press.

Luckman, S. (2012). *Locating cultural work: The politics and poetics of rural, regional and remote creativity*. Palgrave Macmillan.

McRobbie, A. (2015). *Be creative: Making a living in the new cultural industries*. Cambridge: Polity Press.

McRobbie, A., Strutt, D. and Bandinelli, C. (2019). Feminism and the politics of creative labour: Fashion micro-enterprises in London, Berlin and Milan. *Australian Feminist Studies*, 34(100), pp. 131–148.

Merkel, J. (2019). 'Freelance isn't free.' Co-working as a critical urban practice to cope with informality in creative labour markets. *Urban Studies*, 56(3), pp. 526–547.

Mould, O. (2018). *Against creativity*. London: Verso.

Mould, O. and Comunian, R. (2014). Hung, drawn and cultural quartered- rethinking cultural quarter development policy in the UK. *European Planning Studies*, 23(12), pp. 2356–2369.

Oakley, K. and Ward, J. (2018). The art of the good life: Culture and sustainable prosperity. *Cultural Trends*, 27(1), pp. 4–17.

Pasquinelli, C. and Sjöholm, J. (2015). Art and resilience- The spatial practices of making a resilient artistic career in London. *City, Culture and Society*, 6(3), pp. 75–81.

Pine, B.J. and Gilmore, J.H. (2011). *The experience economy*. Cambridge: Harvard Business Press.

Power, D. and Scott, A., eds. (2004). *Cultural industries and the production of culture*. London: Routledge.

Pratt, G. (2010). Collaboration as feminist strategy. *Gender Place & Society*, 17(1), pp. 43–48.

Reimer, S. (2016). 'It's just a very male industry'- gender and work in UK design agencies. *Gender, Place & Culture*, 23(7), pp. 1033–1046.

Schmidt, S. (2019). In the making: Open creative labs as an emerging topic in economic geography? *Geography Compass*, 13(9), p. e12463.

Schumpeter, J. (1959 [1934]). *The theory of economic development: An inquiry into profits, capital, credit, interest, and the business cycle*. Cambridge, MA: Harvard University Press.

Sjöholm, J. and Pasquinelli, C. (2014). Artists' brand building: Towards a spatial perspective. *Arts Marketing: An International Journal*, 4(1/2), pp. 10–24.

# 2 The field of fashion in the digital age

## Insights from global and 'not-so-global' fashion centres

*Taylor Brydges, Marianna d'Ovidio, Mariangela Lavanga, Deborah Leslie, Norma M. Rantisi*

### Introduction

Although fashion is one of the oldest industries, it is also one of the most versatile and innovative. Like most industries, fashion encompasses a range of activities, including design, production, distribution and consumption. As a cultural and creative industry, it is a sector that produces goods valued primarily for their aesthetic rather than utilitarian qualities (Scott, 2000), with its fashion products being subject to rapid change and high levels of market volatility (Entwistle, 2009). These aesthetic qualities are temporally and spatially sensitive, which accounts for the constantly changing form of knowledge that dominates the industry (Entwistle, 2009). Such an unstable form of knowledge depends on a dense constellation of actors in order to be legible. In this chapter, we examine the system of actors that are involved in creating the aesthetics of fashion, or what Bourdieu (1993a) terms 'the field' (see also Hauge and Rykkia, this volume; Nyfeler, this volume) and in doing so, we explore the tension between tradition and innovation (*tension 2*), as well as the tension between isolated and interconnected spaces of creativity (*tension 3*).

In examining such a system, we adopt a dynamic approach that considers the diversity of 'fields' associated with different fashion centres (from the 'global' to the 'not-so-global') and the constellation of sites within each field (including manufacturing, retailing, marketing and consumption). The industry has also been fundamentally transformed by digital technologies, as well as a new range of intermediaries—actors operating between producers and consumers, such as fashion bloggers—which have emerged and are disrupting the old system of production and distribution (Comunian et al., this volume; Crewe, 2017; Brydges and Hracs, 2018; Janssens and Lavanga, 2018; Lavanga, 2018).

DOI: 10.4324/9781003197065-2

Here, we are particularly interested in the implications that recent transformations have for the structure and composition of fields, as well as their power relations and spatiality. As such, in this chapter we seek to contribute to tension 2, by exploring how industrial dynamics in the fashion industry are evolving in the digital age, and to tension 3 by examining the relationship between more isolated or marginal spaces of creativity and leading industry centres (see also Comunian et al., this volume).

The chapter adopts a collaborative and comparative approach utilising data collected from multiple fields of fashion around the world. This mixed-method design includes a textual analysis of international trade publications, newspapers and web materials, as well as over 250 interviews with industry actors in established (such as New York and Milan), and emerging (such as Toronto, Montreal, Stockholm and Amsterdam)[1] centres. Interviews were conducted by the authors with fashion designers (including small independent designers and designers employed by larger fashion houses), as well as other actors in the field, such as those working in fashion education, retail, criticism, public relations, government, trade fairs and so on. The research also draws upon observation in a range of spaces, including boutiques, fashion weeks, trade fairs, exhibitions, studios and manufacturing facilities.

The findings are presented in three parts: an overview of the field of fashion, the impact of digitalisation in challenging established industry norms fashion, and finally, the continuing role of place.

## Literature review: the fashion industry as a socio-spatial 'field'

The fashion industry is a product of a complex array of actors and institutions embedded in place. A key way in which to analyse the industry is through the concept of 'field of cultural production[2]' (Bourdieu, 1993a). According to Bourdieu (1993a), each field operates like a game, with its own rules, forces, players and strategies (Entwistle, 2009). Cultural fields are relatively autonomous from larger fields, structured by economic and political power and class relations, but also influenced by them.

Two main forces are at play in the field: one defined as *autonomous* (the aesthetic considerations not subordinated to economic forces) and one defined as *heteronomous* (economic criteria associated with the market), which structure the field and determine the strong or weak position of creators both within the market, and within the cultural sphere (Bourdieu, 1993b; Entwistle and Rocamora, 2006). As an example, a musician can be very well positioned in the market, producing commercial music, or very special pieces that are appreciated by only a small intellectual circle. This also applies to fashion designers, whose work is subjected to different markets and aesthetic rules.

The fashion designer is dependent upon the work of many different actors: textile and clothes producers, retailers, marketers, consumers, but also those who define aesthetic value, including magazine editors, media critics, bloggers and educators (McRobbie, 2016). As intermediaries, gatekeepers assume a vital importance because they qualify aesthetic commodities and determine their value. Consumers are also an important component of the field, setting patterns of demand and structuring the market.

Fashion also emerges through a global network, consisting of fields located in different places. Although Bourdieu's notion of a cultural field is not an explicitly geographical concept, he does recognise the way in which fields vary across time and space. He cites national variations in cultural fields, and also recognises the importance of geographic capital in establishing the value of artists and their work, with artists coming from places like Paris more likely to move into dominant positions within a cultural field than artists from more provincial centres (Bourdieu, 1993a). Subsequent authors have further developed a spatial conceptualisation of fields, highlighting the bounded nature of many fields (Entwistle, 2009; d'Ovidio, 2015).

As an industry, fashion tends to concentrate in particular cities around the world, known as fashion's world cities. In particular, the industry is dominated by London, Paris, Milan and New York (Kawamura, 2004; McRobbie, 2016; Rantisi, 2002, 2004). These cities hold power in the hierarchy of world fashion cities, and to some degree, they set the trends for the rest of the industry (Larner and Molloy, 2009). This is often attributed to the concentration of international buyers, fashion media and finance present in these cities, which has the effect of consolidating power in a few key locales (d'Ovidio, 2015; Lavanga, 2018; Rantisi, 2004).

A crucial anchoring mechanism of the fashion industry more broadly, and London, Paris, Milan and New York in particular, are fashion weeks. A 'materialisation' of the field of fashion (Entwistle and Rocamora, 2006), such field-configuring events (Lampel and Meyer, 2008) play an important role in bringing the local and global industry together in one place. Fashion weeks also serve as a key site through which new technologies are constituted.

One consequence of the dominance of these cities is that the specificities and attractiveness of local fashion industries outside, but interconnected to, established centres, remain poorly understood. The industry in fashion's world cities has followed a unique, historical trajectory which is difficult for other cities to imitate (Brydges and Hracs, 2019; Larner and Molloy, 2009; Lavanga, 2018; Rantisi, 2004). As such, rather than trying to follow the historical, evolutionary pathways of a world fashion city emerging or 'not-so-global' (Larner, Molloy and Goodrum, 2007) fashion markets must instead find ways to carve out their own trajectories (Brydges, Hracs and Lavanga, 2018; Brydges and Hracs, 2019). For example, in Montreal, this has entailed the development of strong local buyer-designer connections, with independent boutiques and local fashion media giving special recognition to local talent. Montreal designers have also established synergies with other art fields in the city (e.g. graphic design, performing arts) (see Rantisi and Leslie, 2006, 2010; Rantisi, 2011).

Within the European context, we can see examples of tier-two cities that are generating considerable attention. Many of these centres are carving out an alternative identity for themselves. Amsterdam, for example, is becoming the denim capital of Europe (Schuetze, 2012). The city hosts Kingpins, one of the most important denim trade events in the world, and has a flourishing denim ecosystem, including Denim City (a hub for denim innovation), a Jean School (to train future denim experts) and a Denim Days festival (recently exported to New York). The promotion of Amsterdam as a denim capital has been financially backed by the Economic Board Amsterdam and widely promoted in diplomatic visits (Pandolfi, 2015).

While emerging centres have been gaining a lot of traction in recent years, they also struggle to retain talent, often losing designers to larger centres. This is a constant problem in Toronto, where designers often move to New York to pursue their careers (Leslie and Brail, 2011). Not only are there power dynamics within the larger international fashion arena, but within each field, there are also tensions between larger fashion brands and small independent designers (d'Ovidio, 2015; Janssens and Lavanga, 2018; Leslie and Brail, 2011). This has long been evident in New York, where independent design communities have historically emerged in areas outside the established midtown Manhattan fashion district (e.g. SOHO in the 1980s, North of Little Italy in the 1990s, or neighbourhoods in Brooklyn in the 2000s). Within such communities, designers often form their own networks of design, promotion and distribution (e.g. neighbourhood-based events, magazines, markets), and differentiate their products from more corporate players based on more distinctive and artistic styles—styles that eventually get co-opted by the larger brands, heightening tensions within the field (Rantisi, 2002).

Fashion is thus divided into a hierarchy of individual fields, each with its own unique institutional and spatial context. Each centre has a distinct position within the global industry and a distinct visibility within the circuit global fashion events. Larger centres contain different sub-fields. While place is of fundamental importance to the field of fashion, a key question is whether new digital technologies can alter established geographies and power relations.

## The impact of digital technology on fashion fields

Digital technologies have been disrupting many areas of the economy, from transportation to tourism (Martin, 2016). The combination of new digital platforms, artificial intelligence, big data and algorithms are also transforming the fashion industry, creating new systems for value creation and capture (Kenney and Zysman, 2016). These changes have impacted all aspects of the fashion industry, including manufacturing, distribution, marketing, gatekeeping and consumption (Pratt et al., 2012). In this section, we discuss some of these changes and the implications they hold for disrupting established positions within the field of fashion.

Digital technologies are transforming established hierarchies. Much has been said about the ways in which bloggers and influencers have altered the fashion industry, in some cases displacing traditional media institutions, including fashion editors and fashion magazines located in the fashion's world cities (Alexander, 2018). Many of these new actors have been heralded as ushering in a new, more democratic era of fashion, where anyone with a unique viewpoint can make a name for themselves (Esteban-Santos et al., 2018).

However, the democratisation of the industry should not be overstated (Brydges, Hracs and Lavanga, 2018). As social media and digital transformations continue to evolve, there is also evidence to suggest that power is once again becoming consolidated amongst a handful of actors. Indeed, bloggers that were once controversial newcomers, such as Leandra Medine of *Man Repeller* and Chiara Ferragini of *The Blonde Salad*, are today entrenched industry figures in their own right, with multi-million-dollar media and retail empires (Salibian, 2019; Segran, 2019). Once again, emerging and/or independent actors in the industry face challenges in carving out a space in an increasingly competitive industry.

Digital technologies are also reshaping fashion distribution. Consumers are increasingly shifting from malls and high streets to virtual spaces, such as online retailing platforms (Business of Fashion, 2018). As Crewe (2013, p. 775) argues, "online retailing is bringing about transformative shifts in the spaces, times, and practices of fashion consumption . . . The Internet has brought new fashion worlds into the homes, screens, and minds of consumers". One consequence of the rise of digital space is that consumers are interacting with fashion in new ways (Lay, 2018). Some retailers, such as the Gap, are experimenting with augmented reality, allowing customers access to a digital, at-home dressing room. In this format, consumers can try on a larger number of items than in a conventional retail shop (Business of Fashion, 2018). While physical bricks and mortar shops still dominate the majority of fashion purchases, retailers that utilise digital technologies (such as augmented reality, artificial intelligence and large data sets) may be more successful in meeting consumer needs and in product forecasting (Business of Fashion, 2018).

Garments themselves have also gone digital. Amsterdam-based The Fabricant is the first entirely digital fashion house in the world, with the vision to make the city a digital fashion hub. The company produces digital dresses, not meant to be worn in the physical world, but rather worn exclusively in virtual space. In May 2019, The Fabricant sold its first digital dress for $9,500 at an auction in a blockchain summit in New York (interview). Encouraged by students and graduates and teachers, Amsterdam Fashion Institute (AMFI) and its Department of Fashion & Technology have strengthened the school's digital fashion curriculum, experimenting with 3D visual prototyping and virtual reality, and developing a 3D Hypercraft programme.

Another way new technologies are impacting fashion is through blockchain technology, which provides a digital record that helps to better coordinate supply chains, including contracts, invoices and inventory, as well as ethical codes of conduct (Business of Fashion, 2018). This is particularly relevant to large retailers who source products from multiple subcontractors in a variety of countries (Oshri, 2018). The use of blockchain technology may also help companies address logistical and inventory problems within the industry, and better predict demand (Oshri, 2018). This could have the effect of further enhancing the strength of globalised supply chains and reassert the power of large fashion retail brands. Theoretically, however, it could also allow consumers to trace where a product was made and under what conditions, addressing the growing transparency problem and ethical crisis in the industry (Business of Fashion, 2018).

Changes to fashion retailing have also hit one of fashion's strongest institutions hard: fashion weeks. While the long-standing fashion week circuit of New York, London, Milan and Paris continues to dominate the traditional fashion calendar and attract an exclusive set of guests (e.g. fashion editors, fashion buyers, high-profile celebrities), in an era of Instagram and 'see now, buy now' fashion, leading fashion brands and weeks have had to adapt. For example, it is increasingly common to see runway shows that are immediately shoppable and oriented to consumers rather than the fashion establishment (Wood, 2016).

With digitisation, the audience that such shows can reach is also expanding. Many fashion weeks are now live streamed on YouTube and other platforms, in what many have dubbed 'digital fashion weeks', particularly in second tier centres, such as Montreal, which ignited the movement in 2016. Digital Fashion Week, established in 2012, aims to promote industry events for digital designers in both the Global North and South (digitalfashionweek.com), enabling greater exposure for designers outside of a major fashion centre. The shift in the space and scale of fashion weeks in both long-standing and second-tier fashion centres, highlights the importance of examining isolated and interconnected spaces of creativity (tension 3). It also calls attention to the relation, and strain, between more traditional ways of doing fashion week, and new digital approaches (tension 2).

Digital technologies are also altering the marketing of fashion, which is particularly crucial for designers starting their careers. For example, in New York in the 1990s, securing a page in one of the top fashion magazines or trade publications, such as *Women's Wear Daily*, was critical to building brand visibility. Designers in Toronto similarly describe the importance of getting the press to attend their early runway shows, and of luring them with drink tickets (interviews). Fashion designers today operate in a different landscape. In order to "stand out in the crowd" (Hracs, Jakob and Hauge, 2013), designers must be incredibly tech-savvy and position themselves online. Using social media platforms such as Facebook and Instagram provide them with the opportunity to establish their brand and opens new possibilities for smaller players to succeed in a highly competitive sector (Brydges and Hracs, 2018). With these platforms, designers can take consumers backstage, offering relatable content. Digital technologies also allow designers to interact directly with consumers (see also Hauge and Rykkia, this volume) on increasingly personal levels, in contrast to the distant connections facilitated through older intermediaries (e.g. the fashion editor, the buyer or fashion editor). However, the more designers share online, the higher the risk of being copied (Janssens and Lavanga, 2018). Ultimately social media may increase competition between designers.

Contradictory outcomes are thus associated with digital technologies, which offer possibilities for smaller designers, mediators and centres. In the case of designers, they can use the growing array of low-cost technologies as a vital channel to access customers anywhere in the world (Brydges and Hracs, 2018; Crewe, 2017). However, the opportunities of digitalisation come at a cost. Digital platforms require immense investments of emotional and aesthetic labour from those who are working on these platforms (Leslie and Brydges, 2019) and may ultimately strengthen the power of larger players.

## Does place still matter for the contemporary fashion industry?

Reflecting on the scope of these technologies leads one to consider whether geography still matters in the contemporary fashion industry. In particular, our findings suggest that the impact of technology on different fields and phases of the supply chain is reinforcing the role of place. Focusing on the globalising and disembedding effects of digital technologies overshadows the multifaceted and complex dynamics of (re)territorialisation and (re)embedding in the fashion industry. For example, new technologies may ironically facilitate increased networking in physical space. This is the case in Milan, where social networking sites reinforce connections made at industry events and parties (d'Ovidio and Gandini, 2019). In the end, it is the combination of online and off-line relations that is crucial in building the reputation needed to succeed in the field. This is not only the case for designers, but for retailers as well.

Another way that place still matters is in the aesthetic values attached to fashion. Many regions have long specialised in certain segments of the fashion industry, owing to an association with local geographic features (e.g. climate, recreational activities or political sensibilities). Drawing on its local climate and geography, Miami specialises in swimwear, cities in the Pacific Northwest (e.g. Portland or Vancouver) specialise in sports and eco-wear, while Toronto and Montreal are both known for cold weather apparel.

Today, fashion brands use social media to cement these place-based associations, imagined or otherwise (Brydges and Hracs, 2018). For example, in the case of Canada, some brands capitalise on highly specific Canadian imagery (such as the maple leaf, mountains and plaid), while others choose to associate themselves with a more Scandinavian-inspired aesthetic (*ibid*). Over time, a specialisation in a certain aesthetic style is reproduced by the skill sets and infrastructure (technical and social) that develop in a local field.

Place still matters to designers, many of whom choose which city to practice their craft in based on quality of life attributes (Brydges and Hracs, 2019). Many designers stay in Toronto, for example, rather than moving to larger centres like New York, because of the multicultural diversity present in the city, as well as its perceived tolerance and the presence of high-quality schools and health care (Leslie and Brail, 2011). Designers located in Amsterdam stay for similar reasons, relating to its quality of life, international vibe or because they studied there and have friends and family in the city (Wenting, Atzema and Frenken, 2011).

However, fashion design does not seem completely footloose. Even though these location choices seem linked to lifestyle, these cities are hubs to many other creative and non-creative industries, thus suggesting that clustering and agglomeration economies may still play a role (Gong and Hassink, 2017). For example, in the case of Amsterdam, geographical proximity to Paris (ca. 3.5 hours via a high-speed train connection) may explain why many designers prefer living and working in Amsterdam while commuting to Paris when presenting at fashion weeks, showrooms or freelancing for larger fashion houses.

Place also remains central in the realm of distribution and consumption. While digitalisation has brought access to global markets for many smaller independent designers, designers in Toronto and Amsterdam described how consumers still value service, and as a result, a majority of their sales are local, even in a digital era (interviews). In the face of these tensions, it is clear that producers, designers and intermediaries increasingly need to navigate the mix of physical and virtual spaces, understanding their relative advantages and disadvantages. It is also clear that there are contrasting forces at work, facilitating both disembedding and re-embedding.

## Conclusion

Digitalisation continues to shape cultural and creative industries, of which the field of fashion is no exception. Digital technologies have transformed the field of fashion at different scales and in different geographic contexts, with implications for the structure, power relations and spatiality of the field. Drawing on case studies from both established and emerging international fashion centres, we have explored tensions relating to the impact of new digital technologies for industry practices (tension 2) and the divide between isolated and interconnected spaces of creativity (tension 3). We have illustrated that while new technologies can at times work to extend the influence and visibility of established institutions (e.g. Fashion Weeks, or brick and mortar stores), they are also introducing new innovations altogether in retail, marketing and distribution in both peripheral and major fashion centres—a trend that is likely to increase in the context of COVID-19.

And while dominant global fashion centres still command the greatest investments and are most likely to attract talent, digital technologies are allowing 'not-so-global' cities to assert their own identities and networks and to develop niche positions within the global fashion hierarchy. Recent trends suggest that these technologies have the potential to disrupt established hierarchies within the sub-fields of the fashion system and global fashion centres, however, our research indicates that they can also be leveraged to reinforce the power of larger players, such as global firms, that have the capital to acquire and utilise the latest technologies.

Despite the ways in which technologies continue to alter the nature of work for fashion designers, retailers, manufacturers and other actors in the field of fashion, our findings also suggest that place continues to play an important role in the industry. The resources, social connections and historic and aesthetic associations that make a 'place' are also elements that are integrated into different fashion sub-fields.

A key question for the future of digital technology use is how it impacts which resources and capital (cultural and economic) accrue to which places. And, as our cases have been restricted to fashion centres in the Global North, an examination of fashion centres in the Global South could lend further insight into the relation between fashion, geography and digital technologies (Comunian, Hracs and England, 2021). As digitisation is an evolving phenomenon, there is much that remains to be examined, but our chapter has raised some initial questions and insights which may help to inform ongoing analyses of fashion in the digital age.

## Notes

1 Inspired by many discussions at CCE and beyond, we decided to collaborate on this chapter in order to pull together and reflect upon both the commonalities and unique perspectives from our research.

2 Bourdieu included the fashion industry in the field of cultural production, arguing that there is a strong conformity between haute couture and haute culture (Bourdieu, 1993b).

## References

Alexander, E. (2018). *What is the difference between a fashion influencer and blogger? Harper's Bazaar*. Available at: www.harpersbazaar.co.uk/fashion/fashion-news/news/a41898/fashion-influencer-fashion-blogger-definition/.

Bourdieu, P. (1993a). *The field of cultural production*. Oxford: Polity Press.

Bourdieu, P. (1993b). Haute couture and haute culture. In: *Sociology in question, P. Bourdieu*. London: Sage.

Brydges, T. and Hracs, B.J. (2018). Consuming Canada: How fashion firms leverage the landscape to create and communicate brand identities, distinction and values. *Geoforum*, 90, pp. 108–118.

Brydges, T. and Hracs, B.J. (2019). The locational choices and interregional mobilities of creative entrepreneurs within Canada's fashion system. *Regional Studies*, 53(4), pp. 517–527.

Brydges, T., Hracs, B.J. and Lavanga, M. (2018). Evolution versus entrenchment: Debating the impact of digitization, democratization and diffusion in the global fashion industry. *International Journal of Fashion Studies*, 5(2), pp. 365–372.

Business of Fashion. (2018). *The state of fashion*. Available at: https://cdn.businessoffashion.com/reports/The_State_of_Fashion_2019.pdf.

Comunian, R., Hracs, B.J. and England, L. (2021). *Developing creative economies in Africa: Higher education and policy*. London: Routledge.

Crewe, L. (2013). When virtual and material worlds collide: Democratic fashion in the digital age. *Environment and Planning A*, 45(4), pp. 760–780.

Crewe, L. (2017). *The geographies of fashion: Consumption, space, and value*. London: Bloomsbury Publishing.

d'Ovidio, M. (2015). The field of fashion production in Milan: A theoretical discussion and an empirical investigation. *City, Culture and Society*, 6(2), pp. 1–8.

d'Ovidio, M. and Gandini, A. (2019) The functions of social interaction in the knowledge-creative economy: Between co-presence and ICT-mediated social relation. *Sociologica*, 13(1).

Entwistle, J. (2009). *The aesthetic economy of fashion: Markets and value in clothing and modelling*. Oxford: Berg.

Entwistle, J. and Rocamora, A. (2006). The field of fashion materialized: A study of London fashion week. *Sociology*, 40(4), pp. 735–751.

Esteban-Santos, L., Medina, I.G., Carey, L. and Bellido-Pérez, E. (2018). Fashion bloggers: Communication tools for the fashion industry. *Journal of Fashion Marketing and Management: An International Journal*, 22(3): pp. 420–437.

Gong, H. and Hassink, R. (2017). Exploring the clustering of creative industries. *European Planning Studies*, 25(4), pp. 583–600.

Hracs, B.J., Jakob, D. and Hauge, A. (2013). Standing out in the crowd: The rise of exclusivity-based strategies to compete in the contemporary marketplace for music and fashion. *Environment and Planning A*, 45(5), pp. 1144–1161.

Janssens, A. and Lavanga, M. (2018). An expensive, confusing, and ineffective suit of armor: Investigating risks of design piracy and perceptions of the design rights available to emerging fashion designers in the digital age. *Fashion Theory*, pp. 1–32.

Kawamura, Y. (2004). *The Japanese revolution in Paris fashion* (Geral edition). Oxford: Bloomsbury Academic.

Kenney, M. and Zysman, J. (2016). *The rise of the platform economy*. Issues in Science and Technology. Available at: https://issues.org/the-rise-of-the-platform-economy/.

Lampel, J. and Meyer, A.D. (2008). Field-configuring events as structuring mechanisms: How conferences, ceremonies, and trade shows constitute new technologies, industries, and markets. *Journal of Management Studies*, 45(6), pp. 1025–1035.

Larner, W. and Molloy, M. (2009). Globalization, the 'new economy' and working women theorizing from the New Zealand designer fashion industry. *Feminist Theory*, 10(1), pp. 35–59.

Larner, W., Molloy, M. and Goodrum, A. (2007). Globalization, cultural economy, and not-so-global cities: The New Zealand designer fashion industry. *Environment and Planning D: Society and Space*, 25(3), pp. 381–400.

Lavanga, M. (2018). The role of the Pitti Uomo trade fair in the menswear fashion industry. In: R. Blaszczyk and B. Wubs, eds., *The fashion forecasters: A hidden history of color and trend prediction*. London: Bloomsbury Academic Publishing, pp. 191–209.

Lay, R. (2018.). Digital transformation—the ultimate challenge for the fashion industry. *Deloitte*. Available at: https://www2.deloitte.com/ch/en/pages/consumer-industrial-products/articles/ultimate-challenge-fashion-industry-digital-age.html.

Leslie, D. and Brail, S. (2011). The productive role of 'quality of place': A case study of fashion designers in Toronto. *Environment and Planning A*, 43(12).

Leslie, D. and Brydges, T. (2019). Women, aesthetic labour, and retail work: A case study of independent retailers in Toronto. In: L. Nichols, ed., *Working women in Canada: An intersectional approach*. Ontario: Canadian Scholars, pp. 269–286.

Martin, C.J. (2016). The sharing economy: A pathway to sustainability or a nightmarish form of neoliberal capitalism? *Ecological Economics*, 121, pp. 149–159.

McRobbie, A. (2016). *Be creative: Making a living in the new culture industries*. Cambridge and Malden: Polity Press.

Oshri, H. (2018). How technology is shaping the future of the fashion industry. *Forbes*. Available at: www.forbes.com/sites/theyec/2018/10/01/how-technology-is-shaping-the-future-of-the-fashion-industry/#109881241a45.

Pandolfi, V. (2015). *Fashion and the city*. PhD thesis. Uitgeverij Eburon, Delft.

Pratt, A., Borrione, P., Lavanga, M. and D'Ovidio, M. (2012). International change and technological evolution in the fashion industry. In: M. Agnoletti, A. Carandini, and W. Santagata, eds., *Essays and researches. International biennial of cultural and environmental heritage*. Pontedera: Bandecchi & Vivaldi Editori e Stampatori.

Rantisi, N.M. (2002). The local innovation system as a source of "variety": Openness and adaptability in New York City's garment district. *Regional Studies*, 36(6), pp. 587–602.

Rantisi, N.M. (2004). The ascendance of New York fashion. *International Journal of Urban and Regional Research*, 28(1), pp. 86–106.

Rantisi, N.M. (2011). The prospects and perils of creating a viable fashion identity. *Fashion Theory*, 15(2), pp. 259–266.

Rantisi, N.M. and Leslie, D. (2006). Branding the design metropole: The case of Montréal, Canada. *Area*, 38(4), pp. 364–376.

Rantisi, N. M. and Leslie, D. (2010). Materiality and creative production: The case of the mile end neighborhood in Montréal. *Environment and Planning A: Economy and Space*, 42(12), pp. 2824–2841.

Salibian, S. (2019). *Chiara Ferragni awards employees with bonus*. Available at: https://wwd.com/fashion-news/fashion-scoops/chiara-ferragni-employees-bonus-1203147103/.

Schuetze, C. F. (2012). Diplomas for jean genies. *The New York Times*. https://www.nytimes.com/2012/08/21/fashion/21iht-fjeans21.html

Scott, A.J. (2000). *The cultural economy of cities: Essays on the geography of image-producing industries*. London: Sage.

Segran, E. (2019). *Man Repeller's website is the most fun I've ever had online shopping*. Available at: www.fastcompany.com/90365821/man-repellers-website-is-the-most-fun-ive-ever-had-online-shopping.

Wenting, R., Atzema, O. and Frenken, K. (2011). Urban amenities and agglomeration economies? The locational behaviour and economic success of Dutch fashion design entrepreneurs. *Urban Studies*, 48(7), pp. 1333–1352.

Wood, Z. (2016). From catwalk to checkout: How Burberry is trying to reinvent retail. *The Observer*. Available at: www.theguardian.com/business/2016/sep/24/burberry-reinvent-retail-from-catwalk-to-checkout-see-now-buy-now.

# 3 Creative splintering and structural change in Leicester, UK

*Rachel Granger*

## Introduction

Over the last 30 years, the concept of a creative economy has not only entered public consciousness but through a period of consolidation has transformed the urban and economic landscape (Florida, 2002; Landry and Bianchini, 1995; Pratt, 2000). Underpinned by the logic of agglomeration economies and the increasing returns of the new economic geography (Krugman, 1991), creative economies have become synonymous with business concentration in creative clusters (Chapain et al., 2010; Keane, 2011; O'Connor and Gu, 2014), quarters (Evans, 2004; Roodhouse, 2006), and hubs (Dovey et al., 2016).

Against this prevailing logic, this chapter explores the tension between isolated and interconnected spaces of creativity, through a case study of Leicester in the UK. The chapter examines the tensions that arise not only from the changing spaces of creativity within a city but differences between creative sub-sectors. In doing so, the chapter underscores the need for further research on the dynamics of the creative economy beyond popular accounts of world cities and established constructs, but also from aggregated sector cases. This is because from the empirical insights of creative business practice in Leicester, what is gleaned is a sub-national creative cluster under stress and of creative practices extending out of hubs and clusters in ways that are contrary to the literature. While it provides but one view of the creative economy, it suggests that not all creative practice adheres to the same all-encompassing logic, and that conditions for creative business not only can change over time, but operate and are experienced differently at the local and regional scale.

In a departure from the mainstream literature, the chapter is organised around an alternative conceptualisation of the creative economy, which draws on the notion of a creative life cycle. The intention is to create a space for broader discussion on the current state of the creative industries;

DOI: 10.4324/9781003197065-3

something that is only beginning to emerge in critical studies of the last few years. Of broader interest is to position the mainstream rhetoric of the creative industries against competing framings in a way that provokes further review of the creative economy model itself. In the next section, the prevailing view of the creative industries as a cluster and urban manifestation is outlined, and then challenged using Butler's (1980) life-cycle model. At the heart of a life-cycle model is recognition of the changing conditions and behaviour in a city over time, and thus the idea that one creative model is unlikely to endure over time.

## Conceptualising the creative economy

The creative economy is often situated within New Economic Geography (NEG) discourse which provides a framework for understanding the spatial unevenness and organic nature of early creative spaces (O'Connor, 1998). For example, when viewed through the lens of cluster theory (Pratt, 2004; Cooke and Lazeretti, 2008), the creative economy can be framed as the economic efficiencies made possible by geographical proximity and from a local value chain. While this provides an explanatory framework for the structure and practice of a successful creative economy, its 'immediate provenance' and primacy as the growth model for creative economies (see Banks and O'Connor, 2017) restricts wider considerations.

The relational turn or 'second transition' of the New Economic Geography (NEGII) (Bathelt and Gluckler, 2003) counters such a view of the creative industries, by emphasising the inimitability and inherently uneven soft factors in a creative cluster such as relational, social, and aesthetic qualities of economic behaviour. In this approach, the relationship between space and economy is nuanced, so that social space produces a set of conditions, which ultimately drives economic action by shaping the basis for knowledge exchange (e.g. Asheim and Coenen, 2005; Boschma, 2005).

As such, a relational view of creative activities, as a social-learning-innovation nexus, implies inherent discontinuities in the spread and scale of creative work, reflecting different underlying conditions for learning and knowledge exchange. While both a cluster and a relational view of the creative economy offer distinct views of the same landscape, over time the mainstream 'creative cluster' concept has tended to be framed as both unique (based on unique assets) and replicable (universal effect and application) giving rise to serial replication. As a result, the spontaneity of the creative cluster today is 'much less likely' (O'Connor, 2008) and a one-size-fits-all logic prevails (for further consideration of the need to explore the unique dynamics of the creative economy across a range of urban and rural environments, see Capdevila, this volume, and Comunian et al., this volume).

For example, investment in cultural clusters, quarters, and hubs, now occupy a mainstream position in urban regeneration strategies. They reflect the economic value of arts and culture in entrepreneurial cities and the role of the creative economy as an economic fix for depressed economies (Hall, 2000). The European and North American urban creative model (Lazzeretti, Boix and Capone, 2008; Gong and Hassink, 2017) with serial replication in the post-millennium period (Turok, 2003; Lavanga, 2004; McCarthy, 2005), and subsequent adoption in developing economies (e.g. O'Connor and Gu, 2006; Wen, 2012; Murzyn-Kupisz, 2012; Capdevila, this volume) implies a globalisation of a uniform creative model. As tension 3 in this book highlights, there is a need to move beyond the study of 'superstar' cities and consider dynamics of the creative economy in second-tier, regional hubs such as Leicester, as well as in emerging creative economies of the Global South.

It is in this context that further conceptualisation is warranted. While the first years of the new millennium provides a fertile ground for conceptualising the first creative clusters, giving way to a seductive ideology, the urban and economic landscape is now markedly changed and further conceptualisation is important. Understanding the basis for how and where the creative industries model will not only develop, but also thrive (and even sustain) is important for a landscape dotted with serially reproduced creative clusters. There is obvious merit in alternative models that emphasise path-dependency or local viewpoints. Building on Butler (1980), this chapter advocates for the framing of the creative economy as a life-cycle rather than an infinite growth model. Butler's work notes the cyclical nature of discovery, growth, saturation, and then decline phases of a product and host area. In tourism, Butler's work provides an understanding of how tourism resorts grow and decline over time. When applied to the creative economy (Figure 3.1), it provides an alternative framing for creative economic growth that is distinct from the literature and mainstream accounts of infinite growth in star cities.

The creative life-cycle provides a focal point for discussing what leads an industry to change over time; thereby acknowledging that not all creative activity or the areas that host them, are static over time. As such, it entertains the possibility of limitations to growth and market maturity, as well as noting where intervention is needed to reposition a creative activity or space to restore competitiveness. While Butler's model is a generalisation, it provides a broader explanation for growth of areas as early adopters of the creative industries, subsequent consolidation and saturation of a local creative economy, and pathways to eventual decline. In the remainder of this

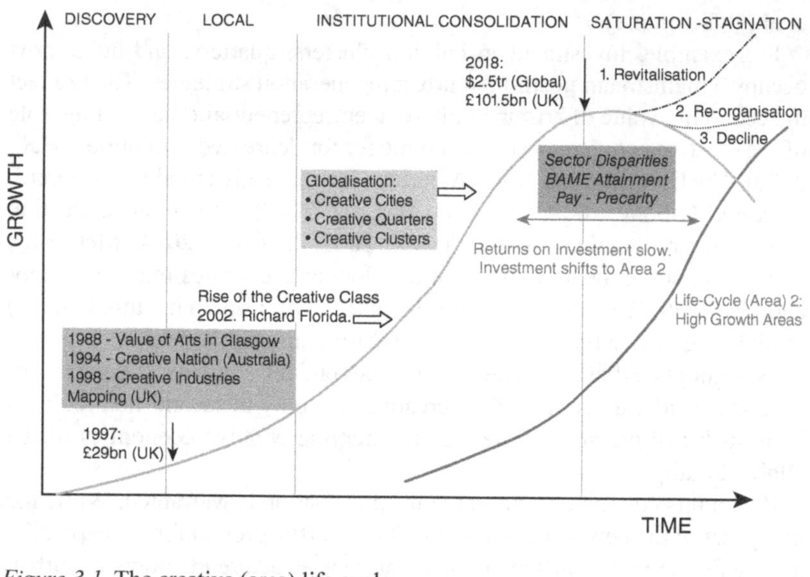


*Figure 3.1* The creative (area) life cycle

Source: Granger (2021), adapted from Butler (1980)

chapter, the creative life-cycle is developed further for use in the creative industries in Leicester, UK.

## Leicester as a creative life-cycle

Leicester is a second-tier city (regional capital) in the East Midlands region of the UK, with a population of 464,000 and a creative population of 34,000 workers and 12,000 businesses (Granger, 2017). During March 2018 to April 2019, face-to-face interviews were conducted with 104 creative workers to gain insights into the City's creative cluster, which here is analysed against Butler's life-cycle model. This has been supplemented by data on 3,518 business interactions in the city captured by a local data platform[1], as well as business data extracted from the UK's Inter-Departmental Business Register (IDBR) held by the UK's Office of National Statistics (ONS). Collectively, this has provided a measure of the creative strength of Leicester vis-à-vis the national (and in particular, London) and international context.

### *Leicester's creative growth*

Leicester's historical significance as a burial site for King Richard III, an international hub for fashion and textiles, and its recent footballing glory, reflect a wider significance of history and museum services, design, and sports excellence, which have further buoyed an expanding cultural and digital media economy, to produce a wide-ranging creative cluster. This has been further supported by Leicester's geography on the transit routes north of London, and significant relocation of London-based studios to premises in Leicester following the 2007/8 recession and in the run-up to Brexit.

Designated investment in Leicester's creative economy emerged from its City of Culture bid in 2010–2013. While arts and culture had been embedded in Leicester since the 1980s to support community cohesion of its diverse population (Leicester was the first minority-majority city in the UK), investment in the commercialisation of arts consumption reflects the way "creativity has been articulated to the neoliberal imaginary" and the broader shift from culture to creativity (Oakley and O'Connor, 2019). Investment in cultural flagships (2007–2014) such as Curve Theatre, Phoenix arts, the LCB creative hub, Makers Yard, and the Cultural Quarter reflects a concerted arts-led regeneration of the city, and coincided with Leicester's attempt to secure the 2017 City of Culture. These early creative investments epitomised the same commercial sentiments of the Blair Government (1997–2001) and early creative ideology (Florida, 2002; Landry and Bianchini, 1995; Pratt, 2000) with a positioning of arts and cultural spaces within a broader entrepreneurial urbanism.

While Leicester's City of Culture bid ultimately failed, the city committed to long-term investment in the arts. A government-supported sector growth plan for Leicester included spending on cultural hubs and venues in the cultural quarter, with investment from the City's two universities (LLEP, 2015). The indirect and induced multiplier effects of early investments supported the city's initial economic recovery, which by 2014 was dominating the East Midlands economy.

### *Leicester's creative splintering*

Despite the early successes, business owners are less sanguine about Leicester's current creative cluster and of possible saturation: "As Nottingham has caught up, it has become more difficult to vie for local cultural consumption" (interview). A network analysis was also conducted on local business interactions extracted from the FLOKK platform in April 2019 (which is presented in Table 3.1). On one level, the high degree of networking among Leicester's creative businesses might be taken to infer strength of the cluster, underpinning local learning and knowledge exchange. However, there are limited interdependencies between university and business sectors that one might expect in a successful creative cluster that supports a *Doing, Using, Interacting* mode of innovation (see Asheim and Coenen, 2005).

*Table 3.1* Interoperability quotients, Leicester (2019)

| | *Sector ii* | | | | | | | | | | | | | | | | |
|---|---|---|---|---|---|---|---|---|---|---|---|---|---|---|---|---|---|
| *(Links from sector i, to sector ii)* | *Creative Economy* | *Digital* | *Smart Economy* | *Construction/ Engineering* | *Transport/Logistics* | *Manufacturing* | *Science/R&D* | *Services* | *Administration* | *Investment* | *Education* | *Health* | *Retail* | *Visitor Economy* | *Community & Third Sector* | *Sub-Total* | *IOQ² (intra)* |
| *Sector i* | | | | | | | | | | | | | | | | | |
| Creative Economy | 496 | 110 | 0 | 4 | 10 | 4 | 4 | 50 | 63 | 48 | 136 | 4 | 30 | 21 | 34 | 1014 | 1.37 |
| Digital | 110 | 132 | 6 | 0 | 2 | 4 | 10 | 46 | 32 | 32 | 58 | 2 | 2 | 2 | 4 | 442 | 0.836 |
| Smart Economy | 16 | 2 | 63 | 0 | 2 | 12 | 6 | 26 | 18 | 31 | 18 | 2 | 0 | 4 | 3 | 203 | 0.8292 |
| Construction/Engineering | 4 | 2 | 6 | 43 | 0 | 6 | 0 | 6 | 16 | 0 | 4 | 0 | 2 | 2 | 2 | 93 | 1.295 |
| Transport/Logistics | 16 | 2 | 0 | 8 | 51 | 4 | 2 | 26 | 17 | 0 | 2 | 0 | 4 | 8 | 4 | 144 | 0.992 |
| Manufacturing | 16 | 12 | 2 | 9 | 8 | 81 | 26 | 10 | 8 | 20 | 16 | 4 | 6 | 8 | 6 | 232 | 0.9779 |
| Science/R&D | 4 | 10 | 0 | 2 | 0 | 2 | 80 | 10 | 4 | 20 | 38 | 0 | 0 | 4 | 0 | 174 | 1.287 |
| Services | 32 | 4 | 6 | 8 | 6 | 9 | 8 | 44 | 24 | 28 | 44 | 4 | 14 | 6 | 10 | 247 | 0.498 |
| Administration | 12 | 2 | 0 | 0 | 2 | 2 | 8 | 6 | 31 | 0 | 47 | 2 | 4 | 14 | 0 | 130 | 0.6679 |
| Investment | 2 | 18 | 0 | 0 | 2 | 2 | 10 | 0 | 4 | 4 | 0 | 0 | 4 | 0 | 8 | 54 | 0.207 |
| Education | 62 | 20 | 14 | 0 | 2 | 6 | 4 | 22 | 44 | 8 | 142 | 0 | 2 | 8 | 21 | 355 | 1.1203 |
| Health | 16 | 2 | 0 | 0 | 0 | 0 | 0 | 4 | 0 | 0 | 6 | 4 | 0 | 0 | 2 | 34 | 0.3295 |
| Retail | 18 | 4 | 4 | 0 | 4 | 4 | 0 | 10 | 6 | 0 | 24 | 0 | 6 | 11 | 0 | 91 | 0.1846 |
| Visitor Economy | 10 | 0 | 0 | 0 | 12 | 0 | 2 | 8 | 12 | 0 | 24 | 0 | 10 | 8 | 2 | 88 | 0.2546 |
| Community & Third Sector | 11 | 2 | 4 | 0 | 0 | 0 | 0 | 4 | 28 | 4 | 44 | 6 | 4 | 18 | 71 | 217 | 0.91644 |
| Sub-total | 846 | 322 | 90 | 74 | 101 | 136 | 160 | 272 | 307 | 195 | 603 | 28 | 88 | 114 | 167 | 3518 | |

Source: Data from FLOKK database

It is not that there is an absence of soft, relational behaviour but that the direction of knowledge transfer is underpinned by intra- rather than inter-sectoral networking, in a way that reveals limits to growth from the city's university research. This is supported by the calculation of interoperability quotients[2] (Table 3.1) that reveal strong networking *within* creative industries ($IOQ^2=1.37$) rather than between university and creative sectors ($IOQ^1=0.72$).

Looking at the broader business picture Leicester's creative learning and networking has become increasingly endogenous, and this is reinforced through local accounts of creative practice. Ninety-three business owners interviewed between 2018–2019, described the way creative knowledge is sourced principally from within the industry, with 87 citing the importance of 'competitors', 'critical friends', and 'peers'. One business leader provides qualification:

> The creative industries are now intensely competitive. First mover advantages have given way to international pressure that require the type of agile thinking and mindsets that are no longer served from the 2- or 3-year research cycles of universities and from informal chat in the local coffee shop or on campus.

Other businesses refer to the tired look and feel of flagship buildings and spaces that were built in the Quarter more than ten years ago:

> There was a time when the LCB was the place to be but it's become a bit forced and a tired model, and while I loved the Design Season, there are more exciting spaces to be outside of the Cultural Quarter.
>
> While I love working at the Makers Yard and being part of the scene . . . the buzz has shifted towards the waterfront and other parts of the Midlands.

These anecdotal accounts of Leicester's creative economy share similarities with Phase 5 Stagnation of the Creative Area Life Cycle. It reflects a weakened role for local universities in the ecosystem, and a deterioration in the appeal of the Cultural Quarter's original features. The relocation of local creative businesses elsewhere in the city, references changing perceptions of the cluster/quarter, with fewer opportunities for interaction, and which over time would undermine the dynamism of the cluster.

## Splintering of the creative industries?

At the national level, there is evidence of a shifting vibrancy of the creative economy, which mirrors some of the local tensions of structural change found in the City of Leicester. Whereas there is evidence of sustained growth overall in the national creative economy (see Table 3.2) in terms of employment, wealth, and exports, which have been framed as key sources of "vibrancy" and "regeneration" in British towns and cities (CBI, 2019), there are also marked differences when disaggregated by sector and spatial scale.

At the national level, there has been contraction in some of the traditional areas of the creative industries such as crafts (-71%), and some sectors such as music and arts, have experienced growth in employment but contraction of GVA. At the same time, there has been remarkable growth in IT and software development (531%), which has been acknowledged more widely: "fast growth partly reflects digital technologies . . . and creative services from advertising, to software, and design" (NESTA, 2018), and the "millennial upswing" has now given way to a technological revolution (Banks and O'Connor, 2017).

At the Leicester level (see Table 3.3), there has been a surge in crafts and design employment particularly in the last five years (326% and 120%, respectively), contraction of visual and performing arts (-7.8%) and continued growth of its digital media sectors. What can be taken from this, is that not all creative sectors perform uniformly and not all creative cities replicate London or the national picture.

Furthermore, while the creative economy has grown overall with net business formation (births) since 2010 at the national and local level (in Leicester), this has not been the case since 2013 (-7.1%). This might be taken to infer early growth giving way to a slowing of the creative economy. This is particularly the case for architecture, crafts, film and TV, publishing, and museums, and to a lesser extent, music, performing and visual arts. At the regional level, too (see Table 3.4), there appears to be a consistent pattern of accelerated creative business formation in the first decade of the new millennium, but a slowing of growth over 2010–2018.

For Leicester (in the East Midlands), creative businesses accounted for 8.1% of all new businesses in 2010 but by 2018, this had decreased to 3.4%. It is not that Leicester's share of creative industries has been lost to other successful cities over this period—the creative industries continue to represent 4% of those nationally—but that the performance of the creative industries relative to the wider economy has changed. While creative business formation was strong in 2010, this is much weaker in 2018.

Overall, what might be taken from these different viewpoints is a longer-term decline in the creative economy nationally particularly in the last five years, against accelerated growth in the wider economy. At the Leicester level, there has been continued growth in the creative economy over the last five years against modest growth in other sectors but this has been marked

*Table 3.2* Longitudinal performance of the creative economy, England

| | *Employment (DCMS Estimates #)* | | | | *Gross Value Added (GVA) (£m) (DCMS #)* | | | | | *Exports in the Creative Economy (£m) (#)* | | | | *Business Births (PA) (IDBR Register*)* | | | | |
|---|---|---|---|---|---|---|---|---|---|---|---|---|---|---|---|---|---|---|
| | *1998* | *2015* | *Change (jobs)* | *Change (%)* | *1997* | *2007* | *2018* | *Change (GVA)* | *Change (%)* | *1996* | *2014* | *Change (Exports)* | *Change (%)* | *2010* | *2013* | *2018* | *Change 2010–18 (%)* | *Change 2013–18 (%)* |
| Advertising and Marketing | 84,900 | 182,000 | 97,100 | 114.37% | 3458 | | | | | 680 | 2771 | 2091 | 307.5% | 3,745 | 3,840 | 3,560 | 4.9% | –0.072917 |
| Architecture | 30,000 | 90,000 | 6,000 | 20.00% | 3100 | 5500 | | 2400 | 77.40% | 380 | 446 | 66 | 17.4% | 1,285 | 2,190 | 1,730 | 34.6% | –21.0% |
| Crafts | 24,200 | 7,000 | –17,200 | –71.07% | | | | | | | | | | 90 | 145 | 120 | 33.3% | –17.2% |
| Design: product, graphic and fashion design | 197,500 | 132,000 | –65,500 | –33.16% | 280 | 480 | | 200 | 71.43% | 350 | 226 | –124 | –35.4% | 2,530 | 3,190 | 2,905 | 14.8% | –8.9% |
| Film, TV, Video, Radio and Photography | 276,006 | 231,000 | –45,006 | –16.31% | 3500 | 4800 | | 1300 | 37.14% | 1210 | 4724 | 3514 | 290.4% | 2,905 | 5,300 | 4,415 | 52.0% | –16.7% |
| IT, Software and Computer Services | 447,000 | 640,000 | 193,000 | 43.18% | 9800 | 28400 | | 18600 | 189.80% | 1400 | 8834 | 7434 | 531% | 14,910 | 22,675 | 21,750 | 45.9% | –4.1% |
| Publishing | 132,000 | 200,000 | 68,000 | 51.52% | 6500 | 10000 | | 3500 | 53.80% | 680 | 2142 | 1462 | 215% | 1,065 | 1,730 | 1,325 | 24.4% | –23.4% |
| Museums, Galleries and Libraries | | 97,000 | 97,000 | | | | | | | | | | | 40 | 95 | 70 | 75.0% | –26.3% |
| Music, Performing and Visual Arts | 166,601 | 286,000 | 119,399 | 71.67% | 6400 | 4000 | | –2400 | –37.50% | 250 | 644 | 394 | 157.6% | 2,990 | 3,515 | 3,775 | 26.3% | 7.4% |
| TOTAL | 1358207 | 2,000,000 | 641,793 | 47.25% | 29700 | 59900 | 101500 | 30200 | 101.68% | 4950 | 19809 | 14837 | 299.73% | 29,555 | 42,685 | 39,650 | 34.2% | -7.1% |

Source: Compiled from DCMS Creative Industries Estimates, IDBR Register, Business Demography (DCMS, 1998, 2017, 2018; ONS, 2010, 2017a, 2017b, 2018)

*Table 3.3* Longitudinal performance of creative economy employment, Leicester

| | *Leicester* | | | | | | | *England* | | | | | | | | | | |
|---|---|---|---|---|---|---|---|---|---|---|---|---|---|---|---|---|---|---|
| | *2015* | *2016* | *2017* | *2018* | *2015* | *Change (jobs) 2015–19* | *Change (%) 2015–19* | *1998* | *2014* | *2015* | *2016* | *2017* | *2018* | *2019* | *Change (jabs) 2014–19* | *Change (%) 2014–19* | *Change (jobs) 1998–2019* | *Change (%) 1998–2019* |
| Advertising and Marketing | 465 | 615 | 405 | 465 | 270 | –195 | –41.9 | 84900 | 182000 | 122020 | 125860 | 112910 | 114185 | 116475 | –65525 | –53.7 | 31575 | 37.2 |
| Architecture | 425 | 550 | 500 | 425 | 500 | 75 | 17.6 | 30000 | 90000 | 44590 | 48320 | 41360 | 51850 | 52940 | –37060 | –83.1 | 22940 | 76.5 |
| Crafts | 1040 | 2125 | 2780 | 1675 | 4435 | 3395 | 326.4 | 24200 | 7000 | 20010 | 19480 | 23115 | 21285 | 25075 | 18075 | 90.3 | 875 | 3.6 |
| Design: product, graphic and fashion design | 3205 | 6450 | 4910 | 3425 | 7060 | 3855 | 120.3 | 197500 | 132000 | 45735 | 53700 | 46180 | 49405 | 48975 | –83025 | –181.5 | –148525 | –75.2 |
| Film, TV, Video, Radio and Photography | 1360 | 1325 | 1420 | 1495 | 1785 | 425 | 31.3 | 276006 | 231000 | 240120 | 241280 | 239750 | 236475 | 240355 | 9355 | 3.9 | –35651 | –12.9 |
| IT, Software and Computer Services | 24S0 | 2535 | 3870 | 4465 | 3435 | 955 | 38.5 | 447000 | 640000 | 406545 | 434060 | 462835 | 464080 | 490755 | –149245 | –36.7 | 43755 | 9.7 |
| Publishing | 165 | 130 | 70 | 45 | 95 | –70 | –42.4 | 132000 | 200000 | 89055 | 86320 | 79675 | 78260 | 73980 | –126020 | –141.5 | –58020 | –43.9 |
| Museums, Galleries and Libraries | 400 | 425 | 140 | 315 | 230 | –170 | –42.5 | | 97000 | 48935 | 53030 | 50120 | 50890 | 59645 | –37355 | –76.3 | –37355 | –38 5 |
| Music, Performing and Visual Arts | 320 | 510 | 350 | 1135 | 295 | –25 | –7.8 | 166601 | 286000 | 63605 | 65030 | 64270 | 58520 | 72380 | –213620 | –335.9 | –94221 | –56.5 |
| TOTAL (Creative Industries) | 9860 | 14665 | 14445 | 13445 | 18105 | 8245 | 83.6 | 1358207 | 2000000 | 1080615 | 1127080 | 1120215 | 1124950 | 1180580 | –819420 | –75.8 | –1120290.8 | –82.5 |
| TOTAL (All Sectors) | 199500 | 205000 | 197500 | 196000 | 203500 | 4000 | 2.0 | | | 13719000 | 13855500 | 13916000 | 13961500 | 14913500 | 14913500 | 108.7 | –13915891 | |

Source: Compiled from IDBR BRES (Nomis, 2020)

*Table 3.4* Longitudinal performance of the creative industries, by UK region

| *New Creative Businesses* | *N. East* | *N. West* | *Yorks & H* | *E. Mids* | *W. Mids* | *East* | *London* | *S. East* | *S. West* | *Wales* | *Scotland* | *N. Ireland* | *UK* |
|---|---|---|---|---|---|---|---|---|---|---|---|---|---|
| 2010 | | | | | | | | | | | | | |
| Total Creative Industries | 490 | 2.085 | 1,400 | 1,155 | 1,515 | 2,670 | 10,645 | 5,420 | 1,930 | 550 | 1,435 | 260 | 29,555 |
| As Percentage of Region | 8.2% | 9.2% | 8.4% | 8.1% | 8.5% | 11.8% | 20.2% | 14.6% | 10.8% | 7.3% | 9.2% | 5.594 | 12.5% |
| As Percentage of UK Creative Industries | 1.66% | 7.05% | 4.74% | 3.91% | 5.13% | 9.03% | 36.02% | 18.34% | 6.53% | 1.86% | 4.86% | 0.88% | |
| All Sectors | 5,975 | 22,705 | 16,630 | 14,325 | 117,805 | 22,580 | 52,755 | 36,915 | 17,835 | 7,505 | 15,530 | 4,590 | 235,145 |
| 2013 | | | | | | | | | | | | | |
| Total Creative Industries | 695 | 3,190 | 1,940 | 1,865 | 2,275 | 3.865 | 15,380 | 7,310 | 3,015 | 860 | 1,920 | 365 | 42,685 |
| As Percentage of Region | 4.3% | 4.2% | 5.4% | 4.3% | 4.0% | 4.1% | 3.7% | 4.2% | 4.3% | 4.7% | 3.9% | 2.794 | 4.1% |
| As Percentage of UK Creative Industries | 1.6% | 7.50% | 4.50% | 4.40% | 5.30% | 9.10% | 36.00% | 17.10% | 7.10% | 2% | 4.5 0% | 0.90% | |
| All Sectors | 9.685 | 35,285 | 23,125 | 22,035 | 25,735 | 32,570 | 83,600 | 50,890 | 25,640 | 11 | 21,540 | 4,855 | 346,275 |
| 2018 | | | | | | | | | | | | | |
| Total Creative Industries | 635 | 2,750 | 1,645 | 1,610 | 2,090 | 3,460 | 15,535 | 6,695 | 2,385 | 725 | 1,720 | 395 | 39,650 |
| As Percentage of Region | 5.3% | 3.6% | 3.6% | 3.4% | 3.6% | 3.3% | 2.9% | 3.2% | 4.8% | 2.8% | 3.8% | 2.594 | 3.3 % |
| As Percentage of UK Creative Industries | 1.6% | 6.9% | 4.2% | 4.1% | 5.3% | 8.7% | 39.2% | 16.9% | 6.0% | 1.8% | 4.3% | 1.0% | |
| All Sectors | 9,440 | 42,975 | 23,405 | 22,850 | 31,300 | 35,880 | 97,300 | 52,795 | 24,100 | 14,045 | 20,585 | 5,900 | 380,580 |

Source: Compiled from IDPR (ONS, 2020)

by sector variations. The data prompts questions about whether we can truly speak of 'one creative cluster' in a given space, and whether aggregated data masks important variations in sector and longitudinal performance, which over time produces differentiation.

Thinking further about the employment experiences of the creative industries, the prevalence of gig working (Hesmondhalgh and Baker, 2010), precarity (McRobbie, 2015) and under-representation of some socio-economic groups in creative sectors (Oakley et al., 2017; Brook, O'Brien and Taylor, 2018; Hesmondhalgh and Saha, 2013) stand at odds with the idea of a strong and successful industry, which has been propagated regionally and nationally. As Brook, O'Brien and Taylor (2018) note, "there has been no golden age of social mobility in the creative industries" and Bakhshi (2019) concedes that there are some longstanding problems that have been neglected by policy makers focusing on headline jobs growth rather than specific sector experiences. These include aspects of low wage, job insecurity, poor management practices, unpaid internships, and socio-economic and ethnic inequality. Several of these aspects are visible in the Leicester labour market.

Whereas London's creative practitioners contribute on average £76,662.31 towards the London economy, this figure is just £12,442.52 in Leicester (Granger, 2017). While this can be explained by Leicester's creative profile skewed towards low-value-adding sectors such as visual and performing arts, design, and routinised IT, it serves to remind us of the non-uniformity of creative practices in different cities.

In Leicester, variations in pay and wealth mix with adverse accounts of creative work:

> The [profit] margins can be slim and in some cases below minimum wage. The situation is getting worse as the labour market expands and more of us compete.

Students also reveal concerns about the strength of the Leicester's creative labour market:

> I'm completing an MA but I just don't see the jobs. . . . Despite all of the hype, . . . I feel I have been sold a career at school, which is now inaccessible.

A successful studio director also notes:

> The industry has sobered up in the last five years . . . For a while it was good with all of the buzzy trendy things but people are no longer drinking

> the Kool-Aid on that. They view things less emotionally, more rationally now. Investors aren't throwing money at creative businesses anymore and there are opportunities to be had in other cities in other countries.

These different accounts reflect a creative city that is not static over time. Florida (2017) concedes that the same forces of neoliberalism that create initial growth can also ravage a city; becoming victims of their own success. Thus, the international map is constantly changing, and a "new generation of cultural hubs is emerging" (Banks, 2019; see Capdevila this volume' Comunian et al., this volume). Consequently, it seems logical to frame creative cities as an evolving narrative rather than a fixed model, where innovation and creative processes are shaped by their environment. While the creative offer in one city consolidates and plateaus, in another city, a new path of discovery may emerge, leading to divergent growth paths.

## Discussion and analysis

While Butler's model is in many ways an over-simplification, the area life-cycle is nevertheless useful in distinguishing between the onset of the creative industries in Leicester (2004–2010), consolidation of creative activities through the City of Culture bid (2010–2013), and the current landscape, which shares similarities with Butler's saturation and stagnation phases. Slowing growth, declining profit, slowing business start-ups, fewer opportunities, and a changing perception of a creative quarter have been interpreted as declining creative performance overall.

While it is certainly not the case that Leicester's creative economy is in terminal decline, and Brexit is a significant factor for more recent performance (of e.g. advertising), there are aspects of Leicester's creative cluster that jar with the wider rhetoric of creative capitalism (Banks, 2020).

Even before the COVID crisis, Oakley and O'Connor (2019, p. 5) observed an apparent interregnum in which the "optimistic threads that marked both cultural and creative industries have unravelled." As they remark, "if the creative economy is going somewhere, it is not clear where and by whom." Against such uncertainty, what the creative (area) life-cycle offers is a new perspective on future possible trajectories of the creative economy, expressed variously in the model as decline, reorganisation, and revitalisation.

### *Decline*

The creative class and cluster theories continue to dominate the creative discourse. Both encapsulate the broad connections between the

commercialisation of culture and ideas in the creative economy and a vibrant post-industrial economy but have evolved into a "new urban crisis" (Florida, 2017). Mobile capital searches continually for new locations and new life-cycles with ever higher returns within a global creative economy. Thus, as Leicester continues to dominate in some creative sectors (e.g. design and digital media), it loses out to new arts and culture in other cities.

### *Reorganisation*

New creative models and formats have begun to emerge in Leicester, which represent a qualitative shift in its creative capitalism. Following a prolonged period in which public stakeholders led on creative flagships for the city, a number of influential developments have set out changes in leadership and investment. These include non-commercial provision of coworking spaces, the creation of practice spaces within social settings such as music venues, libraries, and disused warehouses, and new specialist provision through philanthropy (see Merkel and Suwala, this volume). The introduction of new models, with flexible, open, playful spaces, and diverse funding and ownership arrangements, marks a shift in Leicester's creative economy that reflects the ideologies of alternative economic spaces.

For example, the Leicester Tech Start-Ups Community (network) now operates a large scale accelerator and equity finance model to serve the city; superseding the public-sector programme in the city. This is complemented by Graffwerk, Leicester Interchange, and the Leicester Urban Innovation Lab, which operate their own bootcamp models for grassroots development for arts, innovation, and smart sectors.

As the co-founder of one new development remarks:

> The old model of public-led investment and provision in the creative industries has become overcooked, producing rigidities that have constrained new thinking. Some of the Danish models around informal spaces, flat hierarchies, open rules, and maker mindsets have appealed to us for a number of reasons.

Another leading figure comments:

> The third sector has always been well represented in Leicester over time . . . I don't think it's a stretch to say that organisations like Art-reach, Spark Arts, and the Mighty Creatives are now leading the arts field and responsible for stimulating alternative ways of thinking and working.

### *Digital revitalisation*

Since the summer of 2019, there have been notable additions to the digital areas of Leicester's creative economy. The local growth of mixed reality capability has revitalised some areas of creative production through new consumer audiences and supply chains e.g. VR film making, AR street art, mixed reality performing arts, new types of gaming, and AI creative improvisation.

Public policy and funding support for smart capabilities, including big data, AI, block chain, and robotics has also enabled local companies to diversify in new digital areas. This is particularly apparent in block chain advertising, AI web design, the introduction of robotics in art, new computing platforms, and new chimeras. This has opened up Leicester's creative economy to new investment, consumer interest, and new markets.

## Conclusion

What Butler's area life-cycle model offers the creative discourse is an opportunity to consider the changing nature of local conditions and drivers, which drive the transition from discovery to local development, consolidation, stagnation, and eventually to decline if left unchallenged. It serves to remind us that not all creative cities and clusters operate uniformly, that growth can be finite, and that performance is both conditional and interconnected.

## Notes

1 https://flokk.online—the FLOKK data platform provides a comprehensive view of Leicester's creative activity, capturing the work of a range of businesses operating across a variety of sectors and scales.
2 A measure of the degree of interaction within a sector, as a proportion of intra-sectoral interactions in the region. $IOQ_2 = (Int_{ss}/\Sigma Int_s)/(\Sigma Int_{ss}/\Sigma Int)$ where:

ss = interactions within a sector
s = sector
$\Sigma$Int= Total reference area interactions

## References

Asheim, B. and Coenen, L. (2005). Knowledge bases and regional innovation systems: Comparing Nordic clusters. *Research Policy*, 34(8), pp. 1173–1190.

Bakhshi, H. (2019). *A policy and research agenda for the creative industries, creative industries policy and evidence Centre*, Mar. 25. Available at: www.pec.ac.uk/blog/a-policy-and-research-agenda-for-the-creative-industries.

Banks, L. (2019). The five most creative cities in the world? *BBC*, July 16. Available at: www.bbc.com/culture/article/20190715-the-five-most-creative-cities-in-the-world.

Banks, M. (2020). The work of culture and C-19. *European Journal of Cultural Studies*, 23(4), pp. 648–654.

Banks, M. and O'Connor, J. (2017). Inside the whale (and how to get out of there): Moving on from two decades of creative industries research. *European Journal of Cultural Studies*, 20(6), pp. 637–654.

Bathelt, H. and Gluckler, J. (2003). Towards a relational economic geography. *Journal of Economic Geography*, 3(2), pp. 117–144.

Boschma, R. (2005). Proximity and Innovation: A critical assessment. *Regional Studies*, 39(1), pp. 61–74.

Brook, O., O'Brien, D. and Taylor, M. (2018). Here was no golden age: Social mobility into cultural and creative occupation. *SoccArXiv*, Mar. 27, 2018.

Butler, R.W. (1980). The concept of a tourist-area cycle of evolution and implications for management. *The Canadian Geographer*, 24, pp. 5–12.

CBI. (2019). *UK's creative industries must gleam brightly post-Brexit*. Confederation of British Industry, Oct. 23, 2019. Available at: www.cbi.org.uk/media-centre/articles/uk-creative-industries-crown-must-gleam-brightly-post-brexit-cbi/.

Chapain, C., Cooke, P., De Propis, L., MacNeil, S. and Mateas-Garcia, J. (2010). *Creative clusters and innovation*. London: NESTA.

Cooke, P. and Lazeretti, L. (2008). *Creative cities, cultural clusters and local economic development*. Cheltenham: Edward Elgar.

DCMS. (1998). *The creative industries mapping document, 1998. Department of creative, media and sports*. London: Office for National Statistics.

DCMS. (2017). *DCMS sector estimates 2017: GVA*. London: ONS. Available at: www.gov.uk/government/statistics/dcms-sectors-economic-estimates-2017-gva.

DCMS. (2018). *Creative industries: Focus on employment. Department for culture, media and sport*. London: ONS.

Dovey, J., Pratt, A.C., Moreton, S., Virani, T.E., Merkel, J. and Lansdowne, J. (2016). *The creative hubs report: 2016*. London: British Council.

Evans, G. (2004). Cultural industry quarters: From pre-industrial to post-industrial production. In: D. Bell and M. Jayne, eds., *City of quarters. Urban villages in the contemporary city*. Aldershot: Ashgate, pp. 71–92.

Florida, R. (2002). *The rise of the creative class*. New York: Basic Books.

Florida, R. (2017). *The new urban crisis*. New York: Basic Books.

Gong, H. and Hassink, R. (2017). Exploring the clustering of creative industries. *European Planning Studies*, 25(4), pp. 583–600.

Granger, R.C. (2017). *The sustainability of the creative industries*. Leicester and London: Leicester De Montfort University Centre for Urban Austerity.

Hall, P. (2000). Creative cities and economic development. *Urban Studies*, 37(4), pp. 639–649.

Hesmondhalgh, D. and Baker, S. (2010). *Creative Labour: Media work in three cultural industries*. Abingdon: Routledge.

Hesmondhalgh, D. and Saha, A. (2013). Race, ethnicity and cultural production. *Popular Communication*, 11(3), pp. 179–195.

Keane, M. (2011). *China's new creative clusters. Governance, human capital, and investment*. Abingdon: Routledge.
Krugman, P. (1991). Increasing returns and economic geography. *Journal of Political Economy*, 99(3), pp. 483–497.
Landry, C. and Bianchini, F. (1995). *The creative city*. London: Comedia.
Lavanga, H. (2004). *Creative industries, cultural quarters and urban development: The case studies of Rotterdam and Milan*. Amsterdam: University of Amsterdam.
Lazzeretti, L., Boix, R. and Capone, F. (2008). Do creative industries cluster? Mapping creative local production systems in Italy and Spain. *Industry and Innovation*, 15(5), pp. 549–567.
LLEP. (2015). *Creative industries sector growth plan for Leicester and Leicestershire, Leicester and Leicestershire enterprise partnership*. Available at: www.llep.org.uk/wp-content/uploads/2015/07/LLEP-Creative-Industries-Sector-Growth-Plan1.pdf.
McCarthy, J. (2005). Cultural quarters and regeneration: The case of Wolverhampton. *Planning Practice and Research*, 20(3), pp. 297–311.
McRobbie, A. (2015). *Be creative: Making a living in the new cultural industries*. Bristol: Polity Press.
Murzyn-Kupisz, M. (2012). Cultural quarters as a means of embracing the creative capacity of polish cities. *Quaestiones Geographicae*, 31(4), pp. 63–76.
NESTA. (2018). *Creative nation*. Nesta and the Creative Industries Council. Available at: www.nesta.org.uk/press-release/creative-industries-are-driving-economic-growth-across-the-uk-on-track-to-create-one-million-new-creative-industries-jobs-between-2013-and-2030/.
NOMIS. (2020). *UK business register and employment survey (BRES)/inter-departmental business register (IDBR)*. Office for National Statistics [Accessed 17 Nov. 2020].
Oakley, K., Laurison, D., O'Brien, D. and Friedman, S. (2017). Cultural capital: Arts graduates, spatial inequality, and London's Impact on cultural labour markets. *American Behavioural Scientist*, 61(12), pp. 1510–1531.
Oakley, K. and O'Connor, J. (2019). The cultural industries: An introduction. In: *The Routledge companion to the cultural industries*. Abingdon: Routledge, pp. 1–32.
O'Connor, J. (1998). New cultural intermediaries and the entrepreneurial city. In: T. Hall and P. Hubbard, eds., *The entrepreneurial city: Geographies of politics, regime and representation*. Chichester: John Wiley, pp. 225–240.
O'Connor, J. (2008). *Soft infrastructure, new media and creative clusters: Developing design capacity in China and Australia*. Proposal for Australia Research Council (ARC) Linkage Project (unpublished).
O'Connor, J. and Gu, X. (2006). A new modernity? The arrival of 'creative industries' in China. *International Journal of Cultural Studies*, 9(3), pp. 271–281.
O'Connor, J. and Gu, X. (2014). Creative industry clusters in shanghai: A success story? *International Journal of Cultural Policy*, 20(1), pp. 1–20.
ONS. (2010). *Annual business inquiry, table 1a—gross value added (GVA) of the creative industries, UK*. London: ONS. Available at: www.gov.uk/government/statistics/creative-industries-economic-estimates-february-2010.
ONS. (2017a). *DCMS sector estimates 2017: GVA*. London: ONS. Available at: www.gov.uk/government/statistics/dcms-sectors-economic-estimates-2017-gva.

ONS. (2017b). *UK business: Activity, size and location. Inter departmental business register*. London: ONS. Available at: www.ons.gov.uk/businessindustryandtrade/business/activitysizeandlocation/adhocs/007333analysisoflocalunitsbyinleicestershirebyemploymentsizebandanduksic2007sectionletterprivatesectoronly2016.

ONS. (2018). *Business population estimates, 2018*. London: Office for National Statistics. Table 14, 17. Available at: www.gov.uk/government/statistics/business-population-estimates-2018.

ONS. (2019). *Business Demography 2019*, Table 3.2 count of active enterprises. Inter-departmental Business Register. London: Office for National Statistics

Pratt, A.C. (2000). New media, the new economy and new spaces. *Geoforum*, 31(4), pp. 425–436.

Pratt, A.C. (2004). Creative clusters: Towards the governance of the creative industries production system? *Media International Australia*, 112, pp. 50–66.

Roodhouse, S. (2006). *Cultural quarters. Principles and practices*. Bristol: Intellect.

Turok, I. (2003). Cities, clusters and creative industries: The case of film and television in Scotland. *European Planning Studies*, 11(5), pp. 549–565.

Wen, W. (2012). Scenes, quarters and clusters: The formation and governance of creative places in urban China. *Journal of Cultural Science*, 5(2), pp. 8–27.

# 4 Crowdfunding and co-creation of value

## The case of the fashion brand Linjer

*Anders Rykkja and Atle Hauge*

### Introduction

Crowdfunding happens when an entrepreneur uses a website as a platform to solicit funding directly from the general public (Bannerman, 2013) without the use of traditional financial intermediaries (Mollick, 2014). People working in the cultural and creative industries (CCI) are generally considered early adopters of innovative business models (Dalla Chiesa and Handke, 2020). Indeed, in addition to the perennial problems of access to finance (De Voldere and Zeqo, 2017), digitalisation (Rykkja et al., 2020a) has altered traditional models of investments, value chains and reduced levels of public funding (Boeuf, Darveau and Legoux, 2014). Against this backdrop, this chapter considers the case of Linjer—a unisex bag and watch brand established in 2014—which has been experimenting with crowdfunding as a business strategy. Our analysis highlights how firms navigate digital space and compete in challenging markets such as fashion through strategies linked to space and geography (Rantisi, 2004; Hauge, Malmberg and Power, 2009).

As the bag and watch industry is heavily influenced by history and tradition, the case illustrates the tension between tradition and innovation as sources of value (see also Nyfeler, this volume, on related topic) as well as the tension between individual and collaborative creative practices (see also Brydges et al., this volume Comunian et al., this volume; Granger, this volume). As a newcomer, Linjer negotiates these tensions in part by drawing on their experience in e-commerce to build relations and co-create value with their customers. One of the main findings is that the distinction between the role of consumers and producers becomes ever more blurred in Linjer's business strategy. We demonstrate this by evincing how consumers take part in the production process, both as financiers through crowdfunding, but also as creators of symbolic value. We argue that differential aspects between Linjer and other fashion brands are enabling the customer

DOI: 10.4324/9781003197065-4

to participate in processes of selection, and to an extent, the development of products. This model of ascribing value (Jansson, 2019) succeeds because the underlying value proposition of Linjer communicates a commitment to a production philosophy more so than the product and brand in itself (Currid-Halkett, 2017).

We have conducted a single case study (Flyvbjerg, 2006) using secondary data and netnographic observation as data collection methods (Kozinets, Dolbec and Earley, 2014) where we collected different types of data individually from four main sources. First, we conducted internet searches for blog posts, podcasts and online articles on Linjer and their crowdfunding campaigns. Second, we analysed the content, in the form of posts and their respective comments and community responses, on Linjer's Social Media (SoMe) platforms. We interpret the concept of SoMe as digital platforms and software applications created for the purpose of coordinating networks of people coming together to express ideas, share content and exchange resources and opinions. Third, we extracted and downloaded the entire comment sections, with discussions between Linjer's founders and campaign backers, from the six campaigns promoted on Indiegogo and Kickstarter between 2014 and 2017. Finally, we collected field notes and reflections related to our customer experience shopping with Linjer.

The structure of the chapter is as follows. First, the emerging role of crowdfunding in the fashion industry is described, as well as an overview of fashion firms' behaviour within the crowdfunding economy, involving the understanding of consumers as co-producers. Second, the case of Linjer is outlined. Third, we discuss our findings with a specific focus on strategies and practices highlighting the role of the consumer in this digital ecosystem and the sustainability of crowdfunding as a strategy, as well as discussing future research directions.

## The role of crowdfunding in the fashion industry

E-commerce and other technological changes have a disintermediating effect on retail channels for fashion items, signifying increased competition for money and attention. To overcome such encounters, fashion firms have been early adopters of strategies and innovations to avoid risk and secure profits through the preselection strategies that crowdfunding represents (Crewe, 2013).

Crowdfunding is an alternative financing mechanism to raise financial resources by a promoter (in the role of an entrepreneur) looking to finance a given venture (singular project or ongoing concern). In simple terms, crowdfunding involves using an internet platform to ask people ('the crowd') directly for money. However, there are also other positive effects of running a campaign, mainly immaterial and symbolic. Accordingly, in the context of the creative economy, the mechanism also provides entrepreneurs with a distribution channel, and possibilities to accumulate professional reputation, market visibility, and recognition (Rykkja, Munim and Bonet, 2020b, p. 262).

Even though several investments (equity and loan) and non-investment (reward, donation, and patronage) crowdfunding models exist (De Voldere and Zeqo, 2017), those based on rewards prevail in the creative economy. Reward-based crowdfunding is the exchange of funding for non-monetary rewards, products, or services (Shneor and Munim, 2019), bearing similarities with pre-ordering (Belleflamme, Lambert and Schwienbacher, 2014). About 88% of the estimated 75,000 crowdfunding campaigns in the cultural and creative sector within the EU between 2013 and 2016 used the reward-based model (De Voldere and Zeqo, 2017). In the Nordic countries (Norway, Sweden, Denmark, Finland, and Iceland), data on successful campaigns (1,487) collected for another research project (Rykkja, Munim and Bonet, 2020b) indicate that the fashion industry punched above its weight: despite representing only 7.8% of the total number of campaigns, fashion firms collected nearly a third (29.6%) of the transaction volume. Two features may explain the disproportion. First, the average amount raised per campaign differs between sectors where the design category (including fashion and accessories) is among the highest (De Voldere and Zeqo, 2017). Second, features such as existing production milieus in major agglomerations or tradition for excellence in a given sector (Barbi and Bigelli, 2017; Mollick, 2014) leads to different usage at sectoral level across countries.

Although crowdfunding offers potential financing, the mechanism imposes a different set of demands and skills from the creators and artists involved. As entrepreneurs and campaign promoters, they have to take on the role of marketers and sales departments, forcing them into positions in the value creation process previously performed by other stakeholders (Leyshon et al., 2016). Furthermore, the platform is not a neutral entity; it may choose to act as a gatekeeper, due to its power to grant selected campaigns additional exposure (Davidson, 2019). Thereby, added exposure increases the chances of a given project succeeding, making the platform-as-intermediary an integral part of the value co-creation process. Thus, competition between projects becomes a growing concern for the promoters on these sites. Consequently, it is interesting to analyse Linjer's campaigns to determine what makes them "stand out in the crowd" (Hracs, Jakob and Hauge, 2013).

## The emergence of consumers as co-producers

Increasingly, consumers provide important inputs in the processes of innovation, production, diffusion, and marketing as well as financing through crowdfunding (Grabher and Ibert, 2018). This profoundly changes the former one-way mass communication regime between consumers and producers. The interfaces between the firm and the consumer are turning out to be the locus of value creation and value extraction. To understand this symbiotic relationship between user and producer, Currid Halkett's (2017) theory of the 'aspirational class' is a useful analytical framework. Currid Halkett analyses the signifying of social class through consumption and social practice in the 21st century. One central argument is that conspicuous consumption (Veblen, 1992), Thorstein Veblen's concern hundred years ago, has been democratised. The markers of Veblen's leisure class—revelation of status through consumption of conspicuous goods and an abundance of free time—no longer signify social status or economic wealth because both are widely accessible by a broad swath of the middle classes of western societies (e.g., the transformation of fashion brand Burberry, Hauge and Power, 2013). What defines the aspirational class is an allegiance to a shared set of cultural practices and norms, with acquired knowledge and cultural capital markers of membership rather than economic capital. This is why members of the aspirational class value purchasing organically grown tomatoes directly from a farmer at a local market using a tote or recycled plastic bags as much more symbolically weighted than displaying an oversized Ralph Lauren logo of a horse and rider playing polo as a social practice. Hence, consumption signals a philosophy of life and value system acquired through knowledge and education rather than economic power to buy 'stuff', which is widely available in the marketplace. What is at stake is the prestige and prominence achieved as an outcome of production processes; something Currid Halkett (2017) calls conspicuous production. This explains why using artisanal production methods and being transparent about ethical concerns and the origins of goods become as important as the symbolic value of the brand itself. A process whereby producers (besides providing the actual good) also educate high-end consumers and act as arbiters of taste and culture (Ocejo, 2017). Our findings support this assertion and accentuate Currid Halkett's (2017) argument that this strategy is not for the mainstream market, but rather for a niche of consumers.

Although Linjer is not a fashion firm in the strictest sense, it operates within the parameters of a fashion logic. The economic structure of the fashion industry is in part based on the manufacturing of physical goods (for example, the clothes) with the addition of production of symbolic and aesthetic value (Kawamura, 2005). However, less is known about the practice's firms draw upon to "put fashion into clothes" (Weller, 2004), or in this case in accessories, in a digital context. This is something we will analyse and attempt to explain by investigating how crowdfunding lend itself to processes of value co-creation and production.

This is a question of how value is produced, not solely based on notions of utility but rather on factors that are more immaterial. Most products have a physical and non-physical aspect. The non-physical is often known as the symbolic value of a product (Hauge, 2015). Here, creation of symbolic value is best understood as a socially constructed process, embedded in socio-economic relationships, and governed by feedback through the network and marketplace rather than by distinctive preferences and price signals. Value is thus dependent on contextual conditions and to the dynamics of different actors' interactions (Aspers, 2007). This continuous practice of marketplace positioning, and repositioning resembles a series of *negotiations* (Hauge, 2015). The negotiating processes are social in character and occur in the interaction between various actors. Consumers confirm or negotiate value in these social networks, but meanings are context-dependent and volatile. The evaluation is an active and transaction-intensive process, driven by the mental work of the individual, and accomplished by communities or within creative markets (Suwala, 2014)

In relation to digital co-creation of value, crowdfunding platforms allow consumers to be engaged in new ways. They provide spaces for producers to share and distribute their work and a place for interacting with consumers. This is important because, as argued earlier, value is not an embedded and universal quality but rather created through interactions with customers (Grabher and Ibert, 2018). Specifically, viral marketing via social media and the interactive logic of the Web 2.0 applications implies new principles of transparency and two-way communication. As Leyshon et al. (2016, p. 251) argue, "What was once seen as makeshift alliances and inspired improvisations are gradually settling into a pattern of producer-consumer relations that have the power to redefine what is understood as innovations and markets". Progressively, we see that firms seek to tap into the brutal force of social media, employing users or fans as mediators.

While producers struggle to differentiate and monetise their offerings, some consumers are overwhelmed by the choices and information available to them (Jansson and Hracs, 2018). Finding, evaluating, and choosing what to buy can be so overwhelming that people either tend to choose what they are already acquainted with or avoid making a choice entirely. As a result, many consumers are turning to a range of intermediaries for aid in making sense of the marketplace, signalling a shift in relative importance from creators towards curators of products (Jansson, 2019). Curation is the work carried out by formal (e.g., magazines and guides) and informal (e.g., users of social media platforms) agents in translating and legitimising notions of values in different markets through selection and organisation (Jansson and Hracs, 2018). Thus, the economic support and commentary by users of crowdfunding platforms seen as curatorial work help potential campaign backers make choices in their value negotiations.

## The case of Linjer

Linjer is a fashion brand specialising in leather goods and watches. They started operations in 2014, with a mission to "enable people around the world to upgrade their look without spending an arm and a leg" (Linjer.co, 2014). These ambitions were largely fulfilled through crowdfunding campaigns on Indiegogo (six campaigns) and Kickstarter (four campaigns), where Linjer presold leather bags and watches for nearly $3 million (Linjer.co, 2014, 2015). All the campaigns Linjer have launched, so far, have been a success as they have been able to raise enough money to proceed with the projects. In most cases, the money from backers surpassed that limit by more than 1000%. While Linjer's office relocated to Hong Kong, the location of the company on their campaign pages alternates between Florence, Oslo, and San Francisco. In some cases, relocation was the result of a wish for proximity to be able to oversee sourcing of materials and production (Florence) and logistics (Hong Kong). Most of their products are sold in their online store at a higher price (as an example, $449 retail as opposed to $359 campaign price for the Soft Briefcase) after the crowdfunding campaign is over.

Their products are presented with an emphasis on quality and aesthetics. As argued by Hauge and Power (2013) quality is not a given unity, it is built up by practices, networks, flows and senses: it is constantly 'in the making'. It is a process of qualification-requalification aiming to establish a pattern of characteristics (Callon, Méadel and Rabeharisoa, 2002; Jansson and Waxell, 2011). Linjer makes use of trustworthiness borrowed from both places and other brands. The handbags are presented as "Minimalist Bags Without the Luxury Mark-Up", made with the "finest Italian leather and premium Italian fabrics". These products are sourced in Florence, using the same suppliers as more famous brands, and sold at lower prices by opting for a direct-to-consumer model by leapfrogging the inflated value chain of bigger brands. The watches are presented as Swiss, highlighting the technical information on the performance. Linjer emphasises the quality of the glass, elements of the dial and numbers, and the type of movement (the machinery inside the watch). These are details that speak to watch connoisseurs and indicate quality.

When it comes to aesthetics and design, the 'Scandinavian' heritage is accentuated. The products are described as 'minimalist' and understated. In other words, we see geography and places meet in a mix between the material and immaterial (Hauge and Power, 2013). To sum up, the products are presented as high quality at a low price. Early backers of the campaign are also offered a reduced price. After all—everybody loves a bargain.

## Collective value creation

### *Social media*

Social media is touted by many to be a relatively cost-effective and proficient marketing channel (Brydges and Sjöholm, 2019). Given this fact and Linjer's focus on digital platforms, it is surprising that the brand's SoMe presence can be described as modest at best. Linjer has a profile on all the major SoMe-platforms, but on Twitter, the latest post is from 2016. The activity on Facebook is also meagre. In general, the firm tends to post only when there is a new campaign underway.

Instagram is a favourite for many fashion brands and in particular so-called influencers (Brydges and Sjöholm, 2019). Yet, even here, Linjer keeps a relatively low profile. Their own posts have from 50 to 150 likes, with one peaking at 870. Besides, there are a few other Instagram-profile posts with the hashtag 'linjer' or 'linjerco' (their Instagram profile name). Some of these have some traction with close to four thousand likes. However, when it comes to fashion posts on Instagram, this is still somewhat modest.

On the other hand, Linjer interacts with their fans on SoMe. Most comments receive a response from the company on the same day they are posted. This can either indicate that they do not prioritise these platforms, or that the SoMe strategy is not that fruitful.

### *Online communities*

Online communities, e.g., internet forums based on the exchange of specialised knowledge about a mutually shared topic of interest (Jansson, 2019), are held together by a common interest or goal. SoMe platforms like Facebook are more general in character where the bonds between members are based on more or less personal relations. Online communities can be powerful instruments to influence customers purchasing behaviour. One reason is that online communities allow fashion customers to communicate with each other without any restrictions of time and place (Brogi et al., 2013).

In interviews with Linjer's founder, online forums are accentuated as pulpits to build credibility and reputation as a young brand. In particular, Styleforum (Parr, 2016) a forum dedicated to menswear, has been important in this sense. According to Linjer, a few prominent members discovered the brand, discussed it, and reviewed the products in positive terms. As the forum member, 'Prof. Fabulous' stated to conclude a detailed review of a Linjer briefcase in 2014: "Assuming the final product is as good as the sample I test drove, I'd call it a 'must buy'"(Styleforum, 2014). This review came when Linjer was about to launch their first collection for the market, after a successful crowdfunding campaign. In one of the longest and oldest discussion threads: "The Definitive MANBAG Thread, Part II:2014~", with 721 posts (by May 2019), Linjer is mentioned in several posts. What is noticeable is that even without any formal hierarchy, key members of the Styleforum act as mediators. Due to their expertise and engagement, they enjoyed reverence in that online context. In a status market like this (Aspers, 2007), an endorsement from key forum members is vital. In other words, these individuals act as informal gatekeepers.

### *Campaign pages*

The comment sections of Linjer's campaigns fulfil overlapping functions. At the basic level, it functions as an open customer service chat to resolve shipping and delivery issues. On the next level, comments suggest that an ongoing exchange assesses product characteristics, much the same way as a customer would interact with a shop attendant in physical shops. Although revealing a transparent and direct interaction between Linjer and customers, an important revelation is when communication becomes a form of value negotiation. We can read how disappointed customers criticised Linjer by expressing disappointment over a decision to move assembly of leather bags from Italy to Turkey. The central line of argument was that 'Made in Turkey' or 'China' (the case with other bags), even using Italian linen and leather material, is not the same as 'Made in Italy'. Their perception was that Linjer broke a quality pledge made in their sales pitch, which for some effectively constituted changing the value proposition.

## Negotiating collective value creation

Vargo and Lusch (2008) suggest that firms cannot deliver actual value, but only offer a 'value proposition'. A user creates value by applying and integrating resources included in the value proposition offered by the firm, and the process of value co-creation transpires as a result of this interaction. Crowdfunded fashion goods represent value propositions in its purest form:

without the potential consumers' financial involvement, the product will not see the day of light. Crowdfunding creates a *preselection* market for potential goods that enables well-informed consumers to identify new goods and acquisitions with more ease and at better value (price, style, exclusive origin, identity-forming, and authenticity boosting).

Producers can offer distinctive authenticity by inviting consumers to be a part of the production. Here, crowdfunding becomes a mechanism creating value through an analogous form of co-creation (the act of providing financing) before the point of exchange. Consumers, as users, add resources (pre-ordering), which becomes more than financing (market knowledge as added value) once the information (sales) can be processed. Here, inviting consumers to take part in development processes turn, crowdfunding into a brand development strategy by creating shared meaning and relational bonds between producers and consumers (Foà, 2019), by staging an authentic experience.

Additionally, crowdfunding works as a two-sided strategy to mitigate risk. For the producers, it removes the uncertainty of return on investment. The project will not proceed without sufficient financial backing. For consumers, the safety valve represents a pledge that the producer will deliver on the value proposed. In this way, the crowdfunding platform acts as a guarantor and conduit for value propositions, effectively managing what at project level is structured as a preselection market. This illustrates that the mechanism governing crowdfunding is more than just a fundraising tool.

With a minimalistic design and covert logos, Linjer tap into the ethos of the aspirational class or status market (Currid-Halkett, 2017; Aspers, 2007). In an interview with CNN, Linjer's co-founder Jennifer Chong explains: "If we're asked to choose between cheap or better quality products that will last a long time, we're picking the latter" (Kavilanz, 2016). These aesthetics signal a more subtle form of cultural capital based on information which can often only be accessed or appreciated by a few individuals. Discrete and limited consumption based on informed decisions is the social marker. It is the cost of *information*, not *objects*, that is the main barrier. Consequently, Linjer's modest SoMe activity can be understood in the light of the aspirational class theory. It is the discussions and validation through online communities (like Styleforum) that Linjer strive for rather than 'likes' or diffusion via SoMe (Moser and Moser, 2019). The online presence of Linjer fosters interaction between audiences and the firm. Because of the seemingly social nature of these meetings, we see that the firm engages more directly and 'adds value' through interpersonal interactions with consumers. When empowered and active consumers increasingly participate as co-creators, the meaning of value and the process of value creation is rapidly shifting from a product- and firm-centric view. Rather, the interface between firm and consumer is the locus for value creation. As a result, we are moving towards a situation in which value is the result of an implicit negotiation between consumers and firms (Hauge, 2015).

## Conclusion

One of the most prominent trends we are witnessing in innovation and marketing studies is the involvement of consumers as co-producers of value. This trend underpins the important and evolving tensions between tradition and innovation as sources of value as well as individual and collaborative creative practices discussed throughout the book. Indeed, we see a shifting understanding of value creation—from a more traditional view where the process is integral to the firm or a network of firms, towards an interactive and communicative practice involving active and participating consumers. Increasingly, consumers are becoming an intrinsic part of the digital platforms and valuation ecosystems (Hracs and Webster, 2020). Consumption of symbolically charged goods like fashion, transfer cultural meaning to the individual consumer. At the same time, by consuming these goods, they are also producing value. Value creation is thus best understood as a collective process; as a process fixed in socio-economic relationships and social networks. This is relevant in an economy where signs and symbols are central to value creation. Underpinning this idea is an understanding of an altered and more interconnected relationship between economy and culture.

Linjer has established itself as a profitable and growing business in the crowdfunding economy. They have launched and funded ten campaigns since their start-up in 2014. Still, one could ask if this is a viable and sustainable business model. The crowdfunding market has all the signs of a 'superstar economy', where the winner takes all, and as crowdfunding platforms get larger and less numerous, it may become increasingly difficult to fund projects.

There are several topics and findings in this chapter we think need further research. In particular, we must work to nuance our understanding of innovation in the digital economy by considering the semiotic and symbolic aspects of products. Here, investigations into new power relations and new hierarchies of digital (and others) intermediaries and related changes in customer behaviour seems to be a very promising avenue.

## References

Aspers, P. (2007). Theory, reality, and performativity in markets. *American Journal of Economics and Sociology*, 66, pp. 379–398.

Bannerman, S. (2013). Crowdfunding culture. *Journal of Mobile Media*, 7, pp. 1–30.

Barbi, M. and Bigelli, M. (2017). Crowdfunding practices in and outside the US. *Research in International Business and Finance*, 42, pp. 208–223.

Belleflamme, P., Lambert, T. and Schwienbacher, A. (2014). Crowdfunding: Tapping the right crowd. *Journal of Business Venturing*, 29, pp. 585–609.

Boeuf, B., Darveau, J. and Legoux, R. (2014). Financing creativity: Crowdfunding as a new approach for theatre projects. *International Journal of Arts Management*, 16, pp. 33–48.

Brogi, S., Calabrese, A., Campisi, D., Capece, G., Costa, R. and Di Pillo, F. (2013). The effects of online brand communities on brand equity in the luxury fashion industry. *International Journal of Engineering Business Management*, 5, pp. 5–32.

Brydges, T. and Sjöholm, J. (2019). Becoming a personal style blogger: Changing configurations and spatialities of aesthetic labour in the fashion industry. *International Journal of Cultural Studies*, 22, pp. 119–139.

Callon, M., Méadel, C. and Rabeharisoa, V. (2002). The economy of qualities. *Economy and society*, 31, pp. 194–217.

Crewe, L. (2013). When virtual and material worlds collide: Democratic fashion in the digital age. *Environment and Planning A*, 45, pp. 760–780.

Currid-Halkett, E. (2017). *The sum of small things: A theory of the aspirational class*. Princeton: Princeton University Press.

Dalla Chiesa, C. and Handke, C. (2020). Crowdfunding. In: R. Towse and T. Navarrete Hernández, eds., *Handbook of cultural economics*, 3rd ed. Cheltenham: Edward Elgar Publishing.

Davidson, R. (2019). The role of platforms in fulfilling the potential of crowdfunding as an alternative decentralized arena for cultural financing. *Law & Ethics of Human Rights*, 13, pp. 115–140.

De Voldere, I. and Zeqo, K. (2017). *Crowdfunding—reshaping the crowd's engagement in culture*. Luxembourg: Publications Office of the European Union.

Flyvbjerg, B. (2006). Five misunderstandings about case-study research. *Qualitative Inquiry*, 12, pp. 219–245.

Foà, C. (2019). Crowdfunding cultural projects and networking the value creation. *Arts and the Market*, 9, pp. 235–254.

Grabher, G. and Ibert, O. (2018). Schumpeterian customers? How active users co-create innovations. In: G.L. Clark, M.P. Feldman, M.S. Gertler, and D. Wójcik, eds., *The new Oxford handbook of economic geography*. Oxford: Oxford University Press.

Hauge, A. (2015). Negotiating and producing symbolic value. In: A. Lorentzen, K. Topsø Larsen, and L. Schrøder, eds., *Spatial dynamics in the experience economy*. London: Routledge.

Hauge, A., Malmberg, A. and Power, D. (2009). The spaces and places of Swedish fashion. *European Planning Studies*, 17, pp. 529–547.

Hauge, A. and Power, D. (2013). Quality, difference and regional advantage: The case of the winter sports industry. *European Urban and Regional Studies*, 20, pp. 385–400.

Hracs, B.J., Jakob, D. and Hauge, A. (2013). Standing out in the crowd: The rise of exclusivity-based strategies to compete in the contemporary marketplace for music and fashion. *Environment and Planning A: Economy and Space*, 45, pp. 1144–1161.

Hracs, B.J. and Webster, J. (2020). From selling songs to engineering experiences: Exploring the competitive strategies of music streaming platforms. *Journal of Cultural Economy*, pp. 1–18.

Jansson, J. (2019). The online forum as a digital space of curation. *Geoforum*, 106, pp. 115–124.

Jansson, J. and Hracs, B.J. (2018). Conceptualizing curation in the age of abundance: The case of recorded music. *Environment and Planning A: Economy and Space*, 50, pp. 1602–1625.

Jansson, J. and Waxell, A. (2011). Quality and regional competitiveness. *Environment and Planning A*, 43, pp. 2237–2252.

Kavilanz, P. (2016). *How a leather bag startup hit $1 million in sales in 14 months*. Available at: https://money.cnn.com/2016/10/22/smallbusiness/linjer-bags-jennifer-chong/index.html?sr=twmoney102216linjer-bags-jennifer-chong0703PMVODtopLink&linkId=30236512 [Accessed 1 Nov. 2019].

Kawamura, Y. (2005). *Fashion-ology: An introduction to fashion studies*. Oxford: Berg.

Kozinets, R.V., Dolbec, P.Y. and Earley, A. (2014). Netnographic analysis: Understanding culture through social media data. In: *The Sage handbook of qualitative data analysis*, London: Sage, pp. 262–276.

Leyshon, A., Thrift, N., Crewe, L., French, S. and Webb, P. (2016). Leveraging affect: Mobilizing enthusiasm and the co-production of the musical economy. In: B. J. Hracs, M. Seman, and T. E. Virani, eds., *The production and consumption of music in the digital age*. New York: Routledge.

Linjer.co. (2014). *Linjer's profile and campaign page on Indiegogo*. [online] Available at: www.indiegogo.com/individuals/8736465.

Linjer.co. (2015). *Linjer leather goods*. [online] Available at: www.kickstarter.com/profile/linjerco [Accessed 15 June 2016].

Mollick, E. (2014). The dynamics of crowdfunding: An exploratory study. *Journal of Business Venturing*, 29, pp. 1–16.

Moser, G. and Moser, S. (2019). How the owners of linjer raised $1,000,000 through Kickstarter. *Chasing Foxes*. [online] Available at: www.chasingfoxes.com/owners-linjer-raised-1000000-kickstarter/ 2019.

Ocejo, R.E. (2017). *Masters of craft: Old jobs in the new urban economy*. Princeton: Princeton University Press.

Parr, S. (2016). *This 27-year-old will make $5 million this year selling bags online*. Available at: https://thehustle.co/linjer-bags-revenue [Accessed 12 Nov. 2019].

Rantisi, N. (2004). The designer in the city and the city in the designer. In: D. Power and A. Scott, eds., *Cultural industries and the production of culture*. London: Routledge.

Rykkja, A., Mæhle, N., Munim, Z.H. and Shneor, R. (2020a). Crowdfunding in the cultural industries. In: R. Shneor, L. Zhao, and B.T. Flåten, eds., *Advances in crowdfunding: Research and practice*. Basingstoke: Palgrave Macmillan.

Rykkja, A., Munim, Z.H. and Bonet, L. (2020b). Varieties of cultural crowdfunding. *Baltic Journal of Management*, 15, pp. 261–280.

Shneor, R. and Munim, Z.H. (2019). Reward crowdfunding contribution as planned behaviour: An extended framework. *Journal of Business Research*, 103, pp. 56–70.

Styleforum. (2014). *The RACHford files: Linjer soft briefcase TestDrive*. [online] Available at: www.styleforum.net/threads/the-rachford-files-linjer-soft-briefcase-testdrive.427531/ [Accessed 8 Mar. 2019].

Suwala, L. (2014). *Kreativität, Kultur und Raum: ein wirtschaftsgeographischer Beitrag am Beispiel des kulturellen Kreativitätsprozesses*. Wiesbaden: Springer.

Vargo, S.L. and Lusch, R.F. (2008). Service-dominant logic: Continuing the evolution. *Journal of the Academy of Marketing Science*, 36, pp. 1–10.

Veblen, T. (1992). *The theory of the leisure class*. London: Routledge.

Weller, S. (2004). *Fashion's influence on garment mass production: Knowledge, commodities and the capture of value*. Melbourne: Victoria University.

# 5 Intermediaries, work and creativity in creative and innovative sectors

## The case of Berlin

*Janet Merkel and Lech Suwala*

## Introduction

The organisation of creative and innovative processes has become a crucial issue for workers, companies, organisations and cities in the contemporary cultural economy (Scott, 2008; Krätke, 2011). In contemporary globalised capitalism's search for alternative sources of profit, "new time-space arrangements have to be designed that can act as traps for innovation and invention" (Thrift, 2006, p. 290). Since the early 1980s, business and technology parks provide institutionalised spaces for innovation targeting high-tech start-up and knowledge flows (Brinkhoff, Suwala and Kulke, 2012, 2015). More recently, a variety of new 'performative', flexible shared workspaces with an emphasis on interactivity, sociality and community (e.g. coworking spaces, maker spaces, labs, incubators, accelerators, etc.) have emerged and address freelance workers and young entrepreneurs (Schmidt, Brinks and Brinkhoff, 2014; Merkel, 2017).

In this chapter, we explore for the case of Berlin: the intermediary practices that cluster and network managers in technology parks, hosts of shared workspaces use to enact interactivity, sociality community-building and to create atmospheres and environments conducive to the emergence of new ideas and different forms of working together. These actors have received little attention in academic literature, despite several empirical studies pointing to their crucial role in the socio-economic formation and organisation of institutionalised spaces for innovation and knowledge creation (Brinkhoff, Suwala and Kulke, 2012, 2015) or shared workspaces (Brown, 2017; Jakonen et al., 2017; Merkel, 2019a).

The chapter highlights their intermediary role and practices in establishing social and economic relations and in facilitating a collaborative work atmosphere in which institutions, firms and individual coworkers can engage in collaborative practices and knowledge exchange (see also: Comunian et al., this volume). We conceptualise these practices as

DOI: 10.4324/9781003197065-5

socio-spatial curation situated in institutional embeddings (structures), a specific mode of intermediation that aims at the creation of spaces, situations and affective work atmospheres. By comparing coworking spaces with traditional business and technology parks, we show that community-building initiatives are now a crucial factor in both highly and less formalised environments and that socio-spatial curation has become an important element in the organisation of both techno-innovative and cultural-creative work for future economic development initiatives. The study is based on several empirical research projects conducted in the past ten years in Berlin, Germany.

By interrogating intermediaries' practices and embeddedness, this chapter makes several contributions to the tension between individual and collaborative creative practices. First, it sheds light towards a better understanding of old(er) and new(er) individual and collective ways of working and their underlying organisational structures. Second, it contributes to the discussion about different modes of intermediation (e.g. favourable workplace design, stakeholder orchestration, socio-spatial relationships via facilitating interactivity and sociality) among workers and companies. And third, it provides insights into the interplay between practices and structures of intermediation in diverse innovation and creativity settings. In doing so, it also connects to other chapters in this volume, including Capdevila (Chapter 8) and Granger (Chapter 3) who investigate coworking spaces in Catalonia and Leicester respectively, as well as Comunian et al., (Chapter 9), Jansson and Gavanas (Chapter 10) and Brydges et al., (Chapter 2) who also explore the role of intermediaries in creative processes.

## Intermediaries of work and creativity

### *Intermediaries*

Intermediation is defined "as the process of connecting actors in systems of social, economic, or political relations in order to facilitate access to valued resources" (Stovel and Shaw, 2012, p. 141). Depending on the field, intermediation can be characterised in various ways. In culture and creative industries, intermediaries play a crucial role in connecting creative producers with audiences for their products (Foster and Ocejo, 2015; Maguire and Matthews, 2014). Typically, this involves three different functions: *scanning* and *selection* activities to identify promising talents and emerging trends; shaping the *production* process to ensure a balance between novelty and familiarity of the final product and the *promotion* of the final product towards consumers (Foster and Ocejo, 2015, pp. 6–7).

In high-tech and innovative SME-driven industries, intermediaries, such as cluster or network managers, have also manifold tasks that have been characterised as follows: the *selection* of network/cluster partners, the *allocation* of tasks and resources, the *regulation* of cooperation within the network/cluster and the *evaluation* of individual network relationships, or the entire enterprise network (Sydow and Windeler, 1994, 1997).

However, in both cases, these processes rely on two preconditions. First, intermediaries need to integrate those functions into certain structures that are essential for network, cluster or shared spaces management (Suwala, 2005; Suwala and Oinas, 2012). Second, intermediaries need to allocate meaning towards their practices. In other words, *meaning-making* or *being aware* (Suwala, 2014; O'Connor, 2015; Grabher et al., 2018) is necessary beyond simply *being there* (Gertler, 2003).

Based on this understanding, we highlight a specific mode of intermediation that aims at the creation of spaces, situations and affective work atmospheres. We call this form of intermediation *curation*. In this context, we investigate the role of coworking hosts and cluster and network managers as curators. We find these actors do more than merely "manage" a space or cluster but engage in multiple forms of mediation to facilitate new socio-spatial relationships and create conducive atmospheres. It will be argued that introducing the notion of curation for hosts and managers' practices sheds light on different practices that help to enact coworking spaces or clusters as creative spaces.

### *Work*

Since the 1970s, structural changes in labour markets promote the "individualization of labor in the labor process" (Castells, 2001, p. 282). These changes are fuelled by the expansion of new information and communication technologies (ICT) that enable remote and distributed work practices (e.g. teleworking from home), and change significantly how, when and where people work (Felstead, Jewson and Walters, 2005; Taylor and Luckman, 2018). Against this background, critical cultural and creative labour studies scrutinise the atypical working conditions of independent workers, such as the prevalence of freelancing (Banks, Gill and Taylor, 2013; Bologna, 2018).

In particular, freelancers are exposed to a high degree of risk and entrepreneurial pressure, volatility, flexibility and precariousness in their work. With the spheres of production and reproduction overlapping, the lines of the economic and non-economic sphere and between work and home are blurred for freelance workers (Reuschke, 2016). They must organise their work environment, training and skills development, search for new contracts and work opportunities, and cope with the sequencing of intense work schedules on projects and long off times in-between (Vinodrai and Keddy, 2015). Freelance work has a crucial role in project-based forms of economic production, yet until recently, it was mainly performed at home and invisible (Felstead, Jewson and Walters, 2005; Mould, Vorley and Liu, 2014; Reuschke, 2016).

The recent expansion of 'alternative work arrangements' (Spreitzer, Cameron and Garrett, 2017) has "lead to the emergence and diffusion of new organisational forms and institutions" (Barley and Kunda, 2001, p. 76). We see the rise of public and private institutions, companies and services that cater to the specific needs of these workers and companies with flexible, shared office solutions. These so-called coworking spaces are shared office infrastructures where professionals can rent a desk on a flexible base and are provided with the necessary technical equipment. These workspaces match the financial conditions of freelancers since permanent offices are expensive and do not provide the necessary flexibility in volatile, highly mobile and fragmented labour markets (Merkel, 2015).

Similarly, since the early 1980s, business and technology parks provide—often publicly co-funded—workplaces and laboratories with basic infrastructure (e.g. fast internet, shared meeting facilities) and idiosyncratic machines for companies. These infrastructures are only accessible for pre-selected young innovators or innovation collectives (start-ups) aligning with certain technology fields that have additionally successfully submitted a convincing business plan for a limited time period (Suwala, Kitzmann and Kulke, 2021). Both coworking spaces as well as business and technology parks aim to raise the productivity of their members and to provide stimulating work environments that facilitate creative and innovative processes (Suwala, 2005; Merkel, 2015).

### *Creativity*

Creativity is complex and elusive. Depending on the discipline or purpose of study, definitions around creativity can be formulated in terms of a problem, product, process, person or place (Suwala, 2014). Creativity is possible in all fields of human production that emphasises "creativity as something with meaningful originality, which is useful and valuable at the same time" (Suwala, 2017, p. 95).

Creativity is not restricted to the arts or culture (creation) but also encompasses scientific creativity (discovery), technological creativity (innovation) or economic creativity (entrepreneurship) (Wyszomirski, 2004). In general, it takes a certain minimum level of originality for activities be considered 'creative'. Whereas novelty plays the crucial role within artistic (creation) and scientific creativity (discovery), practical or societal benefit (useful) and economic benefit (valuable) are paramount for technological (innovation) and economic creativity (entrepreneurship). 'Original' refers to the necessary component, 'useful' and 'valuable' to the effectiveness component of creativity (Suwala, 2014, 2017).

These 'varieties of creativity' call for different curatorial practices and work arrangements' of intermediaries (both as individual and collectives), depending on the objectives to either squeeze out, the meaningful originality (novel ideas), the usefulness (putting novel ideas into practice) or the valuable (making profit from the novel ideas).

## Socio-spatial curation of work and creativity

While the notion of curation and the role of the curator has long been associated with the art world, the concept has moved beyond the arts to understand intermediation practices in a range of creative industries (Bhaskar, 2016; Jansson and Hracs, 2018). For example, as a form of cultural intermediation, curation has emerged as a concept to capture practices of value creation that help consumers to select among an increasing abundance of products, services and experiences (Jansson and Hracs, 2018; Pfeufer and Suwala, 2020). And curators are increasingly seen as "a catalyst who prompts dialogue by bringing artists, places and publics together" (Puwar and Sharma, 2012, p. 41).

Following this notion of a curator, we conceptualise *curation* as different practices of mediating social experiences that provide some type of connection, creative engagement, learning or collaborative activity—either in framed institutional spaces of technology parks or rather loose environments of shared workspaces taking the extended understanding of creativity as discussed above (novel, useful, valuable) into account.

Against this background, we conceptualise *curation practices* as a form of socio-spatial curation situated in institutional embeddings, a specific mode of intermediation that aims at the creation of spaces, situations and affective work atmospheres, and that are able to invoke different 'worlds or types of creativity'. The role of intermediaries is to orchestrating places with distinctive qualities by promoting creativity through selection and connection, creation new socio-spatial relationships among many other processes. Our analytical focus will be on the curatorial practices of cluster and network managers and coworking hosts.

## Method

By examining intermediary practices of coworking hosts and cluster and network managers, we shed light on processes that aim at creating atmospheres and environments conducive to cultural-creative or techno-innovative work. We discuss practices of *assembling*, *gathering*, *caring*, *mediating and translating* and *exhibiting and displaying* that help to enact spaces as creative or innovative (Merkel, 2019a). We argue that community-building initiatives are a crucial factor in both formalised and loose environments and that social-spatial curation has become an important element in the organisation of techno-innovative and cultural-creative work for future economic development initiatives.

Our exploration rests on shared interest in practices concerning curation in techno-innovative and cultural-creative work. It compares across several empirical research projects we both have conducted in the past ten years in the particular context of Berlin. Merkel's (2015, 2019a, 2019b) research focuses on work practices of coworking hosts in Berlin's coworking scene. These accounts provide unique insights into practices of social-spatial curation in coworking spaces and the multiple ways mediation facilitates new socio-spatial relationships for cultural-creative workers. In contrast, colleagues and Suwala's (Lange et al., 2011, Brinkhoff, Suwala and Kulke 2012, 2015) investigations concentrate on institutional embeddings and practices of cluster and network managers and their acceleration.

There were a number of common findings in our research that inspired this collaboration. Our accounts emphasise insights towards structures and practices of social and economic curation in more formalised environments such as technology and business incubators or innovation policy measures. We have triangulated our qualitative data sets consisting of interviews, documents and observations. This comprehensive material allows us to gain a better understanding of this mode of cultural and economic intermediation. We also use the comparison between cluster and network managers in business and technology parks and hosts in coworking spaces to understand the widespread application of these intermediation practices in wider urban creative economies. We employ a typology of curatorial practices of coworking hosts inductively derived from an empirical investigation on coworking hosts (see Table 5.1, and in particular, Merkel, 2019a) to heuristically structure our analysis.

The study is a rough and aggregate summary of conducted projects over the years with purpose sampling juxtaposing practices of network/cluster managers and coworking space hosts. It is neither a representative nor

*Table 5.1* Curatorial practices in coworking spaces and innovation centres

| *Curatorial practices* | *Coworking Host* | *Cluster/Network Manager* |
|---|---|---|
| Assembling and arranging | Bringing together people, ideas, objects | Custom tailored technologies, feasibility of business ideas, strategizing on-going tasks |
| Selecting | Members, experts, events | Appropriate firms for technology centres incubators |
| Caring | Well-being of members, effective work atmosphere | Economic objectives, trust and reciprocity among similar firms |
| Mediating | Between coworkers, between coworkers and external experts | "Co-petition" (co-operation and competition) between firms, (in-) formal rules of regulations |
| Meaning making and translating | Valorising freelance and independent work and collaborative workspaces | Sanctions, win-win-situations, Anticipate (mega-)trends |
| Exhibiting and displaying | Coworking space and its spatial design, community, projects and companies | Shared and common visions for trade fairs, PR |

exhaustive sample, but a plausible choice to portray similar, albeit alternating curative practices that have to deal with different varieties of creativity in idiosyncratic institutional arrangements.

## Case studies curatorial practices

### *Case 1: network/cluster managers in Berlin*

Cluster and network managers and their practices are well-documented in studies beginning in the 1990s when the paradigms of 'hybrid /network' organisation (Sydow and Windeler, 1994, 1997) and the 'management of knowledge or innovation' across firms, technology parks or territorial innovation models emerged (Suwala and Dannenberg, 2009; Brinkhoff, Suwala and Kulke, 2012). Cluster and network managers in Berlin were mostly financed by subsidy programmes or public funds with the objective that cluster or network members will realise the potential and necessity of intermediaries in the mid- or long run and finance such positions (Suwala, 2005;

Lerch, 2009). Those positions were highly institutionalised and closely monitored through scheduled activity reports.

Reviewing the curatorial practices laid out in Table 6.1, cluster and network manager practices can be described as follows: *Assembling and arranging* comprises a large strategic potential concerning the controlled input for future activities of the network/cluster. This is a continuous activity closely intertwined with the selection task (Suwala, 2005). The *selection* task goes in line with the question who and what should be included in the cluster/network. In addition to selecting suitable network members with regard to competencies and collective goals, the domain (relevant market, technology or industry) must be defined (Sydow and Windeler, 1997).

Although trust and reciprocity are crucial ingredients in practices of intermediaries (Möllering, 2012), *caring* is first and foremost closely linked to the economic objectives. Here, the issue is the efficient allocation of resources, exemplified by questions about how tasks and resources should be distributed in the network/cluster depending on responsibilities, capital, capacities and resources, and the specific competencies of the companies in the network/cluster (Sydow and Windeler, 1994).

*Mediating* needs to be carried out in negotiations with cooperative and competitive objectives based on balanced partnerships of equals. Moreover, mediating is about regulation: how and by what means the completion of tasks needs to be coordinated? The focus here is the development and implementation of rules within the cooperation (e.g. formal mechanisms such as contracts or informal agreements such as oral agreements). Rules for conflict management are key (Sydow and Windeler, 1997).

Next, the *allocation of meaning* is integrated in practices pertaining to sanctions and win-win-situations. This meaning-making is a *translation* task that needs to be executed by managers towards cluster/network members. It is associated with evaluation, e.g. the how costs and benefits should be determined and distributed in the network/cluster context depending on the performance of each individual network member or the collective success (Brinkhoff, Suwala and Kulke, 2015). Finally, cluster and network managers are responsible for shared and common vision at trade fairs, a mutual appearance and/or marketing at *exhibiting* events (Lange, Power, and Suwala, 2014).

## *Case 2: coworking hosts*

Coworking hosts and their work practices have received little attention, despite several empirical studies pointing to their crucial role in the organisation of coworking spaces (Brown, 2017; Capdevila, 2014). Hosts are often the founders of such spaces (in smaller, more bottom-up coworking spaces) or hired community managers (in more commercially oriented coworking spaces). Often members volunteer as hosts and receive then a reduced membership fee for their community building work.

The motivations of hosts curating coworking spaces are manifold: most aim for cultivating a "sense of community" (Butcher, 2016; Garrett, Spreitzer and Bacevice, 2017) and "meaningful encounters" (Jakonen et al., 2017) among coworkers who often have "different and complementary experiences, skill-sets and contacts, but who share similar values and outlooks" (Brown, 2017, p. 120). Most argue that physical proximity or co-presence alone does not lead to interaction, collaboration, innovation or a lively community in the space (Merkel, 2019a). Rather, coworkers must be engaged in different practices of working together collaboratively. Therefore, hosts also describe their work as "conducting", "mothering", "community-building", or "social gardening" (Merkel, 2015, p. 128).

In *assembling* people, ideas and objects, hosts create distinct spatial settings for work purposes and gatherings and encounters among coworkers. They constantly re-arrange objects within the spaces to adapt to the needs of coworkers but also to surprise. Increasingly, coworking hosts also *select* new coworkers and act as gatekeepers to the professional community in the space. Furthermore, they develop ideas for workshops, educational programs or informal social events to promote shared interests among members (Merkel, 2019a).

Coworking hosts also engage practices of care and in emotional work (Hochschild, 2003 [1983]) by constantly displaying positive emotions for creating welcoming, inclusive and "affective atmospheres" (Gregg, 2017; Pfeufer and Suwala, 2020). Hosts often describe *caring* and hospitality as the most important part of their work. They care for the members' various needs (from personal problems to professional help for work-related projects) and their well-being, the sociable atmosphere in the workspace and the perception of the community outside the space.

Furthermore, in enabling sustained interaction amongst coworkers, as well as with actors outside the coworking space, host actively *mediate* meaningful social relationships. Moreover, as part of their curatorial practices hosts engage in various *meaning-making activities*. Hosts promote positive images of freelance and self-employed work as collaborative and as a social alternative to those highly individualised forms of work (Merkel, 2015, 2019a). For example, the workspace provider WeWork claims to organise a "revolution at work" by building a global "we" generation of knowledge workers (WeWork, 2015). At last, an important part of curatorial practices constitutes the *exhibiting and displaying of* those coworking activities to different publics—either through digital media posts, organised tours or in hosting external events (Beagrie, 2008).

## Discussion

By contrasting the curatorial practices of coworking hosts and cluster and network managers, we have found that both are skilled social actors (Fligstein, 2001) that provide identities, cultural and socio-economic frames to motivate others. Moreover, they have the "ability to induce cooperation in others" (Fligstein, 2001, p. 105). Examining their daily work practices sheds light on a particular mode of how new ideas or practices emerge ('curated serendipity') and how cooperation is enabled in both networks/clusters in technology parks and arrangements in coworking spaces.

Hosts and cluster and network managers bridge and bond companies, people, practices and different types of creativity from different institutional fields to create meaningful encounters among independent workers or businesses (Furnari, 2014; Suwala, 2014). As Foster and Ocejo (2015, p. 13) argue, this type of brokerage is a complex process "involving search, selection, co-production, and tastemaking functions that are accomplished by brokers in multiple formal roles (gatekeeper, coordinator, representative, etc.)." Similarly, we found that both fulfil multiple roles in clusters and coworking spaces (see also: Capdevila, this volume).

Concerning differences in the cases, whereas cluster and network managers operate according to a 'preconceived script' in framed spaces for innovation and entrepreneurship with hands-on and economic rationales, hosts of shared workspaces have to negotiate many forms of creativity in loose or even playful environments, with or without an agenda depending on the type of shared space. Caring seems to be a principal task for hosts of shared workspaces often tied to emotional work and creating affective atmosphere.

On the contrary, cluster and network managers must clearly communicate the rationale or value-added of the work, no matter what they are doing. Important is the distribution of responsibilities, transparency of rules/sanctions, and mutual economic objectives in balanced partnerships of equals among network/cluster members. Hosts, in contrast, are facilitators who connect people and initiate conversations but are not managing potential collaborations.

## Conclusion

In this chapter, we have explored how cluster and network managers and coworking hosts aim to create stimulating, affective and sociable work environments for companies and independent workers. In doing so, we have contributed to exploring the tension between individual and collaborative creative practices (tension 1) and the role of intermediaries in these processes. Results are that intermediary practices span across multiple roles and can differ substantially (depending if individuals or collaborations are addressed). Key is the ability to induce cooperation in others. Hereby cluster and network managers follow a rather 'preconceived script' and coworking hosts often curate social and emotional serendipity.

Transferring our results to the wider context of the field, we confirm that while spatial proximity or 'being there' has long been described as crucial for knowledge sharing and creativity in creative and innovative milieus (Gertler, 2003), shared workplaces, now heralded as a new innovative model for creative or innovative work (Schmidt, Brinks and Brinkhoff, 2014; Wagner and Growe, 2020), are much more about 'being aware' (Suwala, 2014; Grabher et al., 2018), mutual meaning-making, mediation and caring.

In concluding, we want to draw out several points for consideration and potential avenues for further enquiry, many of which were developed during CCE: first, we need more robust empirical investigation into these curatorial practices and their effects. For example, there is little research that combines an interrogation of hosts and cluster and network managers curatorial practices with the experience of workers and companies. Capdevila (this volume) is an example of the type of work that is ongoing research that is needed to advance the field. Second, new methodologies need to be developed that combine the curatorial practices as well as the effects on individual work practices. Third, when examining the curatorial practices of cluster and network managers and coworking spaces the dark side of curation has to be taken into account: the exclusionary tendencies through selection processes and thereby aggravating inequalities in access to these workspaces. Fourth, varieties of creativity require distinctive approaches towards curation depending if creativity (in the narrow sense), innovation and/or entrepreneurship is the main objective of hosts; hereby, the empirical insights from case studies are needed.

## References

Banks, M., Gill, R.C. and Taylor, S. (2013). *Theorizing cultural work: Labour, continuity and change in the cultural and creative industries*. London: Routledge.

Barley, S.R. and Kunda, G. (2001). Bringing work back in. *Organization Science*, 12(1), pp. 76–95.

Beagrie, N. (2008). Digital curation for science, digital libraries, and individuals. *International Journal of Digital Curation*, 1(1), pp. 3–16.

Bhaskar, M. (2016). *Curation: The power of selection in a world of excess*. London: Piatkus.

Bologna, S. (2018). *The rise of the European self-employed workforce*. Milan: Mimesis International.

Brinkhoff, S., Suwala, L. and Kulke, E. (2012). What do you offer? Interlinkages of universities and high-technology companies in science and technology parks in

Berlin and Seville. In: A. Olechnicka, R. Capello, and G. Gorzelak, eds., *Universities- cities- regions*. London: Routledge, pp. 121–146.
Brinkhoff, S., Suwala, L. and Kulke, E. (2015). Managing innovation in 'localities of learning' in Berlin and Seville. In: G. Micek, ed., *Understanding innovation and creativity in emerging economic spaces*. Farmham: Ashgate, pp. 11–31.
Brown, J. (2017). Curating the "third place"? Coworking and the mediation of creativity. *Geoforum*, 82, pp. 112–126.
Butcher, T. (2016). Co-working communities: Sustainability citizenship at work. In: R. Horne, J. Fien, B. B. Beza, and A. Nelson, eds., *Sustainability citizenship in cities: Theory and practice*. London and New York: Routledge, pp. 93–103.
Capdevila, I. (2014). *Different inter-organizational collaboration approaches in coworking spaces in Barcelona*. SSRN. Available at: http://ssrn.com/abstract=2502816.
Castells, M. (2001). *Die Netzwerkgesellschaft*. Opladen: Leske und Budrich.
Felstead, A., Jewson, N. and Walters, S. (2005). The shifting locations of work. *Work, Employment and Society*, 19(2), pp. 415–431.
Fligstein, N. (2001). Social skill and the theory of fields. *Sociological Theory*, 19(2), pp. 105–125.
Foster, P.C. and Ocejo, R.E. (2015). Brokerage, mediation, and social networks. In: C. Jones, M. Lorenzen, and S. Jonathan, eds., *The Oxford handbook of creative industries*. Oxford: Oxford University Press, pp. 405–420.
Furnari, S. (2014a). Interstitial spaces: Microinteraction settings and the genesis of new practices between institutional fields. *Academy of Management Review*, 39(4), pp. 439–462.
Garrett, L.E., Spreitzer, G.M. and Bacevice, P.A. (2017). Co-constructing a sense of community at work: The emergence of community in coworking spaces. *Organization Studies*, 38(6), pp. 821–842.
Gertler, M.S. (2003). Tacit knowledge and the economic geography of context, or the undefinable tacitness of being (there). *Journal of Economic Geography*, 3(1), pp. 75–99.
Grabher, G., Melchior, A., Schiemer, B., Schuessler, E. and Sydow, J. (2018). From being there to being aware: Confronting geographical and sociological imaginations of copresence. *Environment and Planning A*, 50(1), pp. 245–255.
Gregg, M. (2017). *From careers to atmospheres. CAMEo Cuts #3*. Leicester: CAMEo Research Institute for Cultural and Media Economies, University of Leicester.
Hochschild, A.R. (2003 [1983]). *The managed heart: Commercialization of human feeling*. Berkeley and Los Angeles: University of California Press.
Jakonen, M., Kivinen, N., Salovaara, P. and Hirkman, P. (2017). Towards an economy of encounters? A critical study of affectual assemblages in coworking. *Scandinavian Journal of Management*, 33(4), pp. 235–242.
Jansson, J. and Hracs, B.J. (2018). Conceptualizing curation in the age of abundance: The case of recorded music. *Environment and Planning A*, 50(8), pp. 1602–1625.
Krätke, S. (2011). *The creative capital of cities. Interactive knowledge creation and the urbanization economies of innovation*. Oxford: Wiley-Blackwell.
Lange, B., Power, D. and Suwala, L. (2014). Geographies of field-configuring events. *Zeitschrift für Wirtschaftsgeographie*, 58(4), pp. 187–201.

Lange, D., Piesbergen, M., Rohn, K., Schmitz, H. and Suwala, L. (2011). *Nachhaltige Vitalisierung des kreativen Quartiers um den Campus Berlin-Charlottenburg*. Berlin-Adlershof: WISTA Management.

Lerch, F. (2009). *Netzwerkdynamiken im Cluster: Optische Technologien in der Region Berlin-Brandenburg*. Berlin.

Maguire, J.S. and Matthews, J. (2014). *The cultural intermediaries reader*. Los Angeles, London, and New York: Sage.

Merkel, J. (2015). Coworking in the city. *Ephemera: Theory & Politics in Organization*, 15(1), pp. 121–139.

Merkel, J. (2017). Coworking and innovation In: H. Bathelt, P. Cohendet, S. Henn, and L. Simon, eds., *The Elgar compendium to innovation and knowledge generation*. Cheltenham, Northampton: Edward Elgar, pp. 570–588.

Merkel, J. (2019a). Curating strangers. In: R. Gill, A. Pratt, and T. Virani, eds., *Creative hubs in question*. Cham: Palgrave Macmillian, pp. 51–68.

Merkel, J. (2019b). 'Freelance isn't free.' Co-working as a critical urban practice to cope with informality in creative labour markets. *Urban Studies*, 56(3), pp. 526–547.

Möllering, G. (2012). Trusting in art: Calling for empirical trust research in highly creative contexts. *Journal of Trust Research*, 2(2), pp. 203–210.

Mould, O., Vorley, T. and Liu, K. (2014). Invisible creativity? Highlighting the hidden impact of freelancing in London's creative industries. *European Planning Studies*, 22(12), pp. 2436–2455.

O'Connor, J. (2015). Intermediaries and imaginaries in the cultural and creative industries. *Regional Studies*, 49(3), pp. 374–387.

Pfeufer, N. and Suwala, L. (2020). Inwertsetzung von temporären Räumlichkeiten. Standortstrategien von Pop-up-Restaurants in Berlin. *Raumforschung und Raumordnung*, 78(1), pp. 71–87.

Puwar, N. and Sharma, S. (2012). Curating sociology. In: L. Back and N. Puwar, eds., *Live methods*. Malden: Wiley-Blackwell, pp. 40–63.

Reuschke, D. (2016). The importance of housing for self-employment. *Economic Geography*, 92(4), pp. 378–400.

Schmidt, S., Brinks, V. and Brinkhoff, S. (2014). Innovation and creativity labs in Berlin—organizing temporary spatial configurations for innovations. *Zeitschrift für Wirtschaftsgeographie*, 58(4), pp. 232–247.

Scott, A.J. (2008). *Social economy of the metropolis: Cognitive-cultural capitalism and the global resurgence of cities*. Oxford: Oxford University Press.

Spreitzer, G.M., Cameron, L. and Garrett, L. (2017). Alternative work arrangements: Two images of the new world of work. *Annual Review of Organizational Psychology and Organizational Behavior*, 4(1), pp. 473–499.

Stovel, K. and Shaw, L. (2012). Brokerage. *Annual Review of Sociology*, 38, pp. 139–158.

Suwala, L. (2005). *Konzeptionelle Anknüpfungspunkte zur Ausgestaltung von, Good- oder Best-Practices' im interorganisationalen Netzwerkmanagement*. Arbeitspapier am BIEM, FH Potsdam.

Suwala, L. (2014). *Kreativität, Kultur und Raum. Ein wirtschaftsgeographischer Beitrag am Beispiel des kulturellen Kreativitätsprozesses*. Wiesbaden: Springer.

Suwala, L. (2017). On creativity: From conceptual ideas towards a systemic understanding. In: T. Brydges et al., eds., *European colloquium on culture, creativity and economy (CCE) working paper compendium*. Lillehammer: Kunnskapsverket, pp. 82–111.

Suwala, L. and Dannenberg, P. (2009). Cluster-und Innovationspolitik maßgeschneidert. *Standort*, 33(4), pp. 104–112.

Suwala, L., Kitzmann, R. and Kulke, E. (2021). Berlin's manifold strategies towards new commercial and industrial spaces—The different cases of 'Zukunftsorte'. *Urban Planning*, 6(3).

Suwala, L. and Oinas, P. (2012). *Management geography. A conceptual framework*. Management Geography. Available at: www.siemrg.org/images/PDF/4-Oinas-Suwala.pdf.

Sydow, J. and Windeler, A. (1994). Über Netzwerke, virtuelle Integration und Interorganisationsbeziehungen In: J. Sydow and A. Windeler, eds., *Management interorganisationaler Beziehungen—Vertrauen, Kontrolle und Informationstechnik*. Opladen: VS Verlag für Sozialwissenschaften, pp. 1–21.

Sydow, J. and Windeler, A. (1997). Strategisches Management von Unternehmensnetzwerken—Komplexität und Reflexivität. In: G. Ortmann and J. Sydow, eds., *Strategie und Strukturation. Strategisches Management von Unternehmen, Netzwerken und Konzernen*. Wiesbaden: Gabler, pp. 129–142.

Taylor, S. and Luckman, S., eds. (2018). *The new normal of working lives: Critical studies in contemporary work and employment*. London: Palgrave Macmillan.

Thrift, N. (2006). Re-inventing invention: New tendencies in capitalist commodification. *Economy and Society*, 35(2), pp. 279–306.

Vinodrai, T. and Keddy, S. (2015). Projects and project ecologies in creative industries. In: C. Jones, M. Lorenzen, and J. Sapsed, eds., *The Oxford handbook of creative industries*. Oxford: Oxford University Press, pp. 251–268.

Wagner, M. and Growe, A. (2020). Creativity-enhancing work environments: Eventisation through an inspiring work atmosphere in temporary proximity. *Raumforschung und Raumordnung*, 78(1), pp. 53–70.

WeWork. (2015). *WeWork. Create your life's work*. Available at: www.wework.com/ [Accessed 20 Aug. 2015].

Wyszomirski, M.J. (2004). *Defining and developing creative sector initiatives* (Working paper #34). Columbus: The Ohio State University.

# 6 Technology as a source for creativity

## Insights from the Swiss fashion industry

*Judith Nyfeler*

### Introduction

From innovation studies to sociology, fashion has attracted the attention of academics in many fields. The industry is driven by the constant search for novelty (Lipovetsky, 2002), it incorporates the desire for imitation and the need for distinction (Simmel, 2013; Blumer, 1969), it combines order and change (Aspers and Godart, 2013) and it is expected to be both a reproduction and a renewal of itself (Nyfeler, 2019). While high-end fashion such as haute couture claims both uniqueness and craftmanship and has yielded artistic novelties such as Coco Chanel's little black dress or Christian Dior's New Look, the case of ready-to-wear clothing (for the mass market) is based on serial production, imitation and repertoire. Thus, instead of risk it favours 'incremental' technical and aesthetic variations (Crane, 2012; Crane and Bovone, 2006). In these markets the ways in which status signals are produced and reproduced differ. Whereas fashion constantly strives for originality and differentiation from previous collections, mass-market clothing *re*produces, imitates or even copies icons of top players (Aspers, 2010; Doeringer and Crean, 2006; Kawamura, 2005). For both markets, technologies, skill and technical competencies such as methods, processes and practices of manufacturing, seem to be fundamental to claims of quality, status and creativity (see also Rykkja and Hauge, this volume; Brydges et al., this volume). This chapter engages with the tension between tradition and innovation and explores the role of technology. The primary focus is on the technological feasibility of creative ideas and input which are materialised in garments and the tensions found in collaborative creative practices such as defining sources of creativity within organisations.

Treating the creative outcomes of fashion as "stylistic innovations" (Cappetta, Cillo and Ponti, 2006), this chapter implies that variations of cultural meaning and aesthetic alteration are rarely developed from scratch but rather evolve from existing forms. Thus, the case of the fashion industry

DOI: 10.4324/9781003197065-6

not only illustrates disruptive change, but highlights the ongoing relevance of traditional technologies, local crafts and heritage (Ott, 2019). This chapter asserts that creativity is acknowledged and justified by the "intentional configuration of cultural and material elements that is unexpected for a given audience" (Godart, Seong and Phillips, 2020, p. 2). Thus, creativity is understood to be an ascription made by a given audience (Hasse, Mützel and Nyfeler, 2019; Koch et al., 2018). In so doing, the chapter creates a dialogue between work on organisational sociology and creativity within the fashion sector.

In the creative economy, firms generate the main part of their turnover by conceptualising, designing or producing creative goods, services or experiences (Flew, 2013). As a result, they are particularly interested in but also sensitive to the ascription of creativity (Hasse and Nyfeler, 2019). This can pose a problem as the value of novelty and its usefulness is unclear in advance of production. Moreover, the creative industries are characterised by volatile market dynamics and uncertain consumer behaviour (Caves, 2002; Dowd, 2004). Under these circumstances, it is helpful to orient activities and decisions towards imprinted and established forms of technologies such as the well-known knitting or sewing machine. Indeed, organisations combine established and well-known technologies with the (partial) renewal of models, and through this approach manage to reduce everyday uncertainty and complex decision-making (Lampel, Lant and Shamsie, 2000; Mora, 2006). This raises the question of what role technology plays in the making of creativity?

To address this question, two fashion firms, which cover two variants of fashion design, were selected and studied. Both will remain anonymous. Despite differences related to their technologies, product ranges, location, formal structure (departments, contracts, business form) as well as audiences and consumers respectively (see also Nyfeler, 2019), both firms rely highly on their technological competencies which they preserve rather than change or renew. Hence, this chapter shows the relevance of similarities in manufacturing garments and how the role of technology can be problematic to creativity. The first company in this study is a couture fashion house that originally traded with silk fabrics and then started to produce a variety of fashionable goods such as ceramics, shoes, textile and wooden toys, jewellery and interior products. Only a few products are produced in their Zurich headquarters, most are produced abroad by trusted suppliers. The other company is a knitted clothing producer specialising in the "complete knitting technique". This is a 3D-knitting computer by which the producer can reduce textile waste and sewing costs—making production in Swiss facilities viable. Both firms were founded in Switzerland and are headquartered there.

In order to be able to keep up with the fashion system's temporality and the subsequent time restrictions, these firms avoid technological change and instead pursue a strategy of technological preservation and stylistic tradition. Technology is regarded as a source of creativity. Although often considered a major constraint in creative processes, it can also orient creative actions and enable the making of new fashionable clothing (Cardinale, 2018). The effect of constraints for creativity have been studied in various realms: norms and conventions (Becker, 2008), networks (Patriotta and Hirsch, 2016), team processes (Rosso, 2014), practices (Ortmann and Sydow, 2018), organisational structures (Chen, 2012) and resources (Sonenshein, 2014). In this study technology limits artistic freedom, yet, in turn, allows for creativity within technological boundaries. The aim is to investigate the tension between innovation and tradition in the fashion sector and expand the discussion of the significance of technology in the creative industries by highlighting the ways in which know-how and technical skills interact with creativity.

## Technology: materialised techniques and tacit knowledge

With *Organizations in Action*, James D. Thompson (2008 [1967]) laid a foundation for organisational sociology for which technology plays a particularly important role. Similarly, Henry Mintzberg (1989) conceptualised an organisational configuration (the adhocracy) which responds to environmental dynamics while buffering the technological core of the organisation from almost any outside influence and disruption. Capturing technology as the product of human action and stressing its assumed structural properties, Orlikowski (1992) discusses the dialectical instead of deterministic relation of technology and organisations. In his study on crafts, sociologist Richard Sennett (2010) considers the craftsman, the tools and the skills and how these elements are not mutually exclusive, but interrelated and mutually stimulating. These approaches emphasise the duality of technology that consists of technical aspects, for example, infrastructure, instruments, elaborate equipment and methods of production, knowledge, such as expertise, (professional) skills, (practical) experience and competencies alike. Often, such knowledge is tacit, personal, action-based and seldom reflected upon (Ray, 2009; Tsoukas, 2003). Technology is thus essentially social: collaborations with social network contacts (producers, suppliers, collaborating partners) are the 'vessel' in which this kind of technology is made accessible. Hence, technology is key to the (technical and cognitive) structure of an organisation and its core competencies that consist of all activities and decisions. Still, and due to buffering, technology can adapt to changes in the environment. From a technology research perspective, technologies are being defined as "a set of pieces of knowledge both directly 'practical' (related to concrete problems and devices) and 'theoretical' (but practically applicable although not necessarily already applied), know-how, methods, procedures, experience of successes and failure and also, of course, physical devices and equipment" (Dosi G., 1982, p. 151).

Integrating assumptions and principles presented above, I conceptualise technology as any materialised equipment or techniques (instruments, tools, machines, infrastructure) and tacit knowledge (know-how, experience, skill, practices) that is relevant for the design, development and manufacture of creativity in a fashion organisation. Following this definition, technologies in fashion range from 'hardware' such as knitting or sewing machines, computers, looms, screen printing instruments, pattern making technology and plotters, to 'software', such as crochet and embroidery skills, the experience of seamstresses and the skilful practices and (shared) know-how of social network contacts with suppliers and producers. Substituting a technology may change the productive focus of an organisation, how it responds to changing demands or addresses other clientele. Hence, technological necessities determine moments of creativity in the fashion sector.

## Research design

On the basis of multi-sited ethnographic field work (Czarniawska, 2014; Flyvbjerg, 2006), the case of fashion was studied in two Swiss fashion companies. As illustrated in Table 6.1, ten semi-structured interviews were conducted with a range of actors (Liebold and Trinczek, 2009; Kuhlmann, 2005).

Participant observation in the form of shadowing (Czarniawska, 2007) was also conducted over the course of four years from 2014–2018. Furthermore, illustrative documents such as newsletters, corporate magazines and advertisements were taken into account to investigate self-description and external perception (Ventresca and Mohr, 2005; Froschauer, 2009). The analysis was conducted using an abductive interpretation (Tavory and Timmermans, 2014), which entails moving between empirical material, theoretical concepts and possible explanations in order to theorise the social phenomenon.

Firm One, originally founded in 1894 by former silk traders, is a Zurich-based company taken over by the current owner family in 1974. The new owner diversified the range of products in order to establish a solid economic base and to be able to further develop and test new product ideas. The former designer (originally educated as a teacher) now holds the position as Art Director and CEO. Her daughter, the current designer, has no formal

*Table 6.1* List of empirical data material

| *Data material: Overview* | *Company One* | *Company Two* |
|---|---|---|
| Product range | Silk goods, cashmere, scarves, women's wear, ceramics, house interior, jewellery, perfume, tea and accessories | Knitted men's and women's wear |
| Interviewees | In total on site: 5<br>Design/Creative Director (2x);<br>Art Director/CEO;<br>Production Manager;<br>PR Manager;<br>HR & Store Manager | In total on site: 4<br>CEO;<br>Design/Creative Director;<br>Web Manager/Design Assistant;<br>Machine Engineer;<br>Programmer |
| Shadowing sequences | Fashion show;<br>presentation sketches AW16;<br>first sample meeting;<br>second sample meeting;<br>internship four weeks;<br>fitting for fashion show;<br>presentation sketches SS17;<br>designer meets couturier;<br>photo shooting | Photo shooting;<br>sample meeting underwear;<br>annual convention Swiss fashion association/ textile fair;<br>presentation sketches SS16;<br>expert interview with designer in school;<br>head of shops meeting;<br>recap meeting w/ designer;<br>producer's meeting at production site;<br>season kick-off |
| Additional document material | Price tags and brand labels, newsletter, collection booklet, invitation to collaborative events/ season openings, fashion show leaflet, correspondence w/ producers | Newsletter, collection booklet, brand labels, advertisements, sketches of drafts (incl. technical descriptions) |

training in fashion design, but was raised in the "colourful and inspiring environment" of the workshop and store, as the members of the organisation repeatedly mentioned. The designer and her assistant, the product manager, PR manager, Art Director and the Sales and HR manager represent the design team. The central technology of this company is the experience and expertise of their partners (meaning their producers and suppliers). Their knowledge covers a vast range of techniques and practices, such as hand-embroidery, crocheting, classical sartorial artistry, hand rolling of silk scarves or screen printing. Thus, the product range is extensive as are its fundamental practices and the base of its partnerships. Products include scarves of several fabric qualities, jewellery and accessories, shoes women's clothing, perfumes, tea, ceramics, interiors, furniture and toys. The company focuses less on one particular technique or method of production, but rather tries to absorb all technological possibilities provided by the producers and suppliers. Due to their large range of products, they are considered a generalist. Although mainly produced in India and Hungary, the company positions its products within a rather high price segment (approximately €300 for a knitted sweater in 2020).

Firm Two was founded in 1993 by the designer herself. After several changes in management, the firm was taken over by the knitting manufacturer in 2013. The designer is formally trained in fashion design with experience from the French couture system. The design team is composed of the designer, CEO, production manager and the machine engineer, who also translates the stylistic ideas into the programmable IT language of the knitting computer. Due to the technological focus on a very particular 3D seamless knitting machine called "complete knitting technology", the firm concentrates on knitted goods for women and men only. The technology involves specific roles such as a skilled machine engineers and programmers who translate item sketches into the language of the computer. The designer is bound by contract to produce her garments on these machines despite the fact that the manufacturer owns "normal" knitting machines, too. This is due to the unique selling point the machines offer: production is cheaper and the garments are entirely produced in Switzerland. Also, it produces less waste and is considered more sustainable. The firm is thus a specialist producer and by combining Swiss production with the 'complete technology' approach it is able to sell the knitted products within a mid-price segment (approximately €150 for a knitted sweater in 2020).

## Technologies in fashion: the knitting machine and social partnerships

The fashion firms depend heavily on the technical equipment and skills of manufacturers. Several examples from the field illustrate the process of how the generation of novel designs adapt to technology. As mentioned, technology not only involves machine-driven techniques, infrastructure, machines or instruments but also social contacts such as long-term partnerships with producers and suppliers or collaborations with artists. These contacts use tacit knowledge, crafts, skills and the practical experience of producers on the basis of which fashion design is realised. I will now show how techniques orient feasibility, sketching considerations and compromises, and how knowledge and expertise inform choices of craft.

One of my first field contacts stressed the importance of technological fit. In one conversation, the CEO pointed to the importance of the alignment between any design task and the knitting machine:

> In the discussion with the designer, the colours are defined, the cut is elaborated, and the stitches are discussed with help from the knitting developer. Then we agree upon the feasibility of the machine. This means: feasible or not? Actually, there is no other question more relevant than this one.
>
> (Field notes, 3.2.2015)

This comment stresses that "feasibility", in this case the technological possibilities to realise the designer's ideas, is central to the fabrication of novel models and styles. Despite the machine-focus, the "feasibility" is estimated by the programmer who also effects the content of the future collection. The interpretation of the knitting programmer has an effect on the development and production of the garment, notably on decisions of pattern, weight, density, stitches and elasticity. Technology is considered during the whole process—from the first sketch until the garment is sold. Here, the designer explains how the specific knitting technology accompanies her in practically all considerations and how it frames the processes of ideation, creation and production:

> One part of the design task is how to approach designing new knitwear. Techniques and material are set from the beginning and affect the sketching process. Another part of the task is to limit the programming efforts that are required in order to realise the new idea. Therefore, she considers the competencies of the knitting developer and the computer from the start. The more effort in pattern and knitting development, the more work for the programmer.
>
> (Field notes, 30.6.2015)

While this description of the designer's task depicts the idea of the designer as a problem solver who needs to review all options, later, the designer also feels subordinated to the technological restraints. In her store, she explains which objects turned out as she wanted, and where this was not possible due to technical limitations:

> Another example are holes in a design similar to a Japanese textile design. According to her sketches, the holes should be of different

> sizes; this was simply ignored, and all holes have been adjusted to the similar size. Now, the design looks rather unsophisticated and less original. . . . In terms of the choice of the colours, the designer already considered the options of the 3D-knitting computer: Because the thread passes longitudinally, no pattern can be knitted. This is the reason why if there is a pattern, it is a single stripe design, often in a single colour.
> (Field notes, 17.12.2015)

As the designer lacks core knowledge of the knitting computer program, she has a hard time assessing whether the knitting developer lacks skill, knowledge or time to develop a particular style. Other considerations include the capability of the machine itself and budgetary constraints laid out by the CEO. Thus, the central position of technology has as much effect on the development as on the production of the fashion collection. Yet, negotiating technological prerequisites can also be an energising challenge. The following quote indicates how established relationships with partners affects processes of creation, particularly choices of craft:

> Ok—now another example, . . . of our embroiderers, because this [embroidery] is done mostly by men, upto 99% Muslim men from India. Their wives are often simple housewives, of whom most are extremely technically skilled, so now we started to crochet with them. . . . the relationship started because they were our silk suppliers, but over the course of time he has become our partner and his daughter is a friend of mine. . . . And this relation has become so close that he boldly asked: 'Hey! It would be cool if we could crochet something again. Do you have any idea?' I do like it quite a lot and I appreciate that, so I usually say: 'Well, yes, I think, I might be coming up with an idea. . .' And then they have something to do. You know, I think actually this is very important.
> (Interview with designer, 30.10.2014)

This quote reveals that knowledge can be a source of inspiration. Further, it illustrates that technologies in the form of partnerships can inform processes of creativity and therefore, partnerships are a fundamental part of technology. In this case, the competence and know-how such as traditional skills of crafts or experience in sartorial workmanship inspires and influences the development and the production of fashion collections. There have been times where the design team recognises that the seamstresses have not carried out a task as they were supposed to:

> One skirt has a frayed waistband. Apparently, the classically trained seamstresses do not like such details. When the design team visits

> the seamstresses in Hungary, the seamstresses spoke so long about a detail until they have departed completely from the original idea. The designer and her assistant have to ponder and balance, how important the original idea was and if there is a compromise they would agree upon.
>
> (Field notes, 5.11.2015)

Compromises and trade-offs are a common way to agree upon diverse ideas. In this respect technology enables and stimulates new ideas and provides it with technological fit with the design. The firm responds to the request of the partner and subsequently creates space for the support and long-term maintenance of its partnerships.

To work together with long-term partners offers several advantages, from trust and mutual interests, to shared past experience (one of the designers calls it “mental cloud”) and collaborative learning. Social network contacts affect and contribute to the realisation of new sketches and designs as much as machines do. Often, the participants mentioned their “partners”, sometimes called “friendships”, as the relevant cause why they do business, also from a perspective of loyalty and being a reliable business partner. Reliability refers to the importance of trust in the relationship, social responsibility and reduced transaction costs. This commitment is as important for technological reasons as these contacts provide the firm with relevant knowledge and skills. On these grounds, and as it contains its core technological resource, the company has an interest in maintaining long-term relationships. Therefore, the organisations strive to maintain a loyal and trustworthy relationship with their suppliers.

Finally, from the first presentation of the new collection meeting to the final viewing of the last prototypes before final production, it can be summarised that the discussions in the design team (between the designer, her assistant, the production manager and the machine engineer) prepare the fit with the central technology. Often, the results are compromises such as adjustments to the style or hemline of a garment, elasticity, the choice of raw material thread or fabric quality) used, or sizing. These examples illustrate the fashion firm’s pathway to success: they build on their imprinted choices of core technologies in order to avoid additional costs. This is followed by adopting alternatives (inspirations, trends) from outside the organisational boundaries and adapting them to the specific technological language.

## Conclusion

This chapter sought to understand how technology interacts with novelty. Drawing on examples from two Swiss fashion firms, I explored the role that technology plays in the making of creativity, or the outcomes that are perceived to be creative. Taken together, the findings attest to the interdependence of technology and creativity and they nuance our understanding of the tension between tradition and innovation. The empirical material shows that by building a technological foundation on the basis of production methods (specific competencies of machines, instrument equipment and infrastructures and human and social relations, knowledge and competencies), the two fashion firms found a reliable way to constantly renew their collections without having to renew the production processes. Specifically, by re-interpreting existing models (cuts, colours, patterns) and preserving methods (ways of production, craft techniques) that have proven valuable, they create novel collections that match the style of the firm. While the renewed fashion is communicated to the market, the installed technological core seldom is. In that respect, the organisation manages to buffer the operational core from dynamic and constantly changing alternatives influenced by the industry such as fashion weeks, magazines or trend forecasts (see Brydges et al. this volume). Hence, the willingness to substantially change approaches, technologies and production methods is reduced. Therefore, technology constrains the infinite possibilities of creativity and renewal deriving from the fashion field and in so doing it operationalises possible options that are in turn adopted and interpreted according to technological feasibility. Approaching the making of novel fashionable clothing from a technology perspective, the resulting technological familiarity does not hinder the production of new styles, but endows the outcome with legitimacy, understanding as well as acceptance and thus also accounts for the organisation's economic survival (Cappetta, Cillo and Ponti, 2006).

The chapter has also explored issues related to the tension between individual and collaborative creative practices including specific steps along the production process. Technologies have been demonstrated to be socio-technical 'actors' that are crucial creative processes (Hracs and Webster, 2020). As a result, the chapter not only looked at interactions between a range of human actors, including designers and seamstresses but included human based and non-human technologies such as 3D-knitting machines or traditional embroidery techniques. Moreover, key decisions in the creative process were highlighted including the choice and integration of machine or human labour, the technical translation of sketches into programmable designs or the interpretation of the technical pattern by the classically trained seamstresses. These examples emphasise the influence of technology—techniques *and* knowledge—on the development of novel fashionable garments and the realisation of whole collections within the fashion industry.

To conclude, this chapter contributes to current debates in creativity and organisation literature in three ways: first, it *relocates* the focus of constraints as a generative input for creativity; secondly, it *reconfigures* the importance of techniques and knowledge for the making of creativity in fashion collections; and third, it *reveals* the interdependence of technology and creativity as well as tradition and renewal. Thus, this study encourages us to question the negative impact of limits on creativity and to instead see constraints as prolific sources of inspiration and novelty.

## References


Aspers, P. (2010). *Status and standard markets in the global garment industry. MPIfG Discussion Paper*, 05/10. Cologne: Max Planck Institute for the Study of Societies.

Aspers, P. and Godart, F.C. (2013). Sociology of fashion: Order and change. *Annual Review of Sociology*, 39, pp. 171–192.

Becker, H.S. (2008). *Art worlds*, 2nd ed. Berkley and Los Angeles: University of California Press.

Blumer, H. (1969). Fashion: From class differentiation to collective selection. *The Sociological Quarterly*, 10(3), pp. 275–291.

Cappetta, R., Cillo, P. and Ponti, A. (2006). Convergent designs in fine fashion: An evolutionary model for stylistic innovation. *Research Policy*, 35(9), pp. 1273–1290.

Cardinale, I. (2018). Beyond constraining and enabling: Toward new microfoundations for institutional theory. *Academy of Management Review*, 43(1), pp. 132–155.

Caves, R.E. (2002). *Creative industries: Contracts between art and commerce*, 2nd ed. Cambridge, MA: Harvard University Press.

Chen, K.K. (2012). Organizing creativity: Enabling creative output, process, and organizing practices. *Sociology Compass*, 6(8), pp. 624–643.

Crane, D. (2012). Globalization, organizational size, and innovation in the french luxury fashion industry: Production of culture theory revisited. In: B. Morean and A. Alacovska, eds., *Creative industries: Critical readings. Volume 3: Organization*. London and New York: Berg, pp. 80–102.

Crane, D. and Bovone, L. (2006). Approaches to material culture: The sociology of fashion and clothing. *Poetics*, 34, pp. 319–333.

Czarniawska, B. (2007). *Shadowing and other techniques for doing fieldwork in modern societies*. Malmö: Liber AB.

Czarniawska, B. (2014). *Social science research: From field to desk*. London and Thousand Oaks: Sage.

Doeringer, P. and Crean, S. (2006). Can fast fashion save the US apparel industry? *Socio-Economic Review*, 4(3), pp. 353–377.

Dosi, G. (1982). Technological paradigms and technological trajectories. *Research Policy*, 11, pp. 147–162.

Dowd, T. (2004). The embeddedness of cultural industries. *Poetics*, 32, pp. 1–3.

Flew, T. (2013). *Global creative industries*. Cambridge: Polity Press.

Flyvbjerg, B. (2006). Five misunderstandings about case-study research. *Qualitative Inquiry*, 12(2), pp. 219–245.

Froschauer, U. (2009). Artefaktanalyse. In: S. Kühl, P. Strodtholz, and A. Taffertshofer, eds., *Handbuch Methoden der Organisationsforschung*. Wiesbaden: VS Verlag für Sozialwissenschaften, pp. 326–347.

Godart, F., Seong, S. and Phillips, D. (2020). The sociology of creativity: Elements, structures, and audiences. *Annual Review of Sociology*, 46(14), pp. 1–22.

Hasse, R., Mützel, S. and Nyfeler, J. (2019). Hauptsache neu? Die Organisation von Innovation und Kreativität. In: M. Apelt, I. Bode, R. Hasse, U. Meyer, V. V. Groddeck, M. Wilkesmann, and A. Windeler, eds., *Handbuch Organisationssoziologie*. Wiesbaden: Springer.

Hasse, R. and Nyfeler, J. (2019). Alles nur Mode? Organisation und Sprache in den Creative Industries. *Zeitschrift für Soziologie*, 48(5–6), pp. 401–417.

Hracs, B.J. and Webster, J. (2020). From selling songs to engineering experiences: Exploring the competitive strategies of music streaming platforms. *Journal of Cultural Economy*, 14(2), pp. 240–257

Kawamura, Y. (2005). *Fashion-ology*. London and New York: Bloomsbury Academic.

Koch, J, Wenzel, M., Senf, N.N. and Maibier, C. (2018). Organizational creativity as an attributional process: The case of haute cuisine. *Organization Studies*, 39(2–3), pp. 251–270.

Kuhlmann, M. (2005). Beobachtungsinterview. In: S. Kühl, P. Strodtholz, and A. Taffertshofer, eds., *Handbuch Methoden der Organisationsforschung*. Wiesbaden: VS Verlag für Sozialwissenschaften, pp. 78–100.

Lampel, J., Lant, T. and Shamsie, J. (2000). Balancing act: Learning from organizing practices in cultural industries. *Organization Science*, 11(3), pp. 263–269.

Liebold, R. and Trinczek, R. (2009). Experteninterview. In: S. Kühl, P. Strodtholz, and A. Taffertshofer, eds., *Handbuch Methoden der Organisationsforschung*. Wiesbaden: VS Verlag für Sozialwissenschaften, pp. 32–56.

Lipovetsky, G. (2002). *The empire of fashion*. Princeton, NJ: Princeton University Press.

Mintzberg, H. (1989). *Mintzberg on management: Inside our strange world of organizations*. New York: The Free Press.

Mora, E. (2006). Collective production of creativity in the Italian fashion system. *Poetics*, 34(6), pp. 334–353.

Nyfeler, J. (2019). *Die Fabrikation von Kreativität Untertitel: Organisation und Kommunikation in der Modeindustrie*. Bielefeld: Transcript.

Orlikowski, W.J. (1992). The duality of technology: Rethinking the concept of technology in organizations. *Organization Science*, 3(3), pp. 398–427.

Ortmann, G. and Sydow, J. (2018). Dancing in chains: Creative practices in/of organizations. *Organization Studies*, 39(7), pp. 899–921.

Ott, R. (2019). The cordwainer's lair: Contingency in bespoke shoemaking. In: E. Bell, G. Mangia, S. Taylor, and M.L. Toraldo, eds., *The organization of craft work. Identities, meanings, and materiality*. Abingdon and New York: Routledge, pp. 196–216.

Patriotta, G. and Hirsch, P.M. (2016). Mainstreaming innovation in art worlds: Cooperative links, conventions and amphibious artists. *Organization Studies*, 37(6), pp. 867–887.

Ray, T. (2009). Rethinking Polanyi's concept of tacit knowledge: From personal knowing to imagined institutions. *Minerva*, 47(1), pp. 75–92.

Rosso, B.D. (2014). Creativity and constraints: Exploring the role of constraints in the creative processes of research and development teams. *Organization Studies*, 35(4), pp. 551–585.

Sennett, R. (2010). *Handwerk*. Berlin: Berlin Taschenbuch Verlag.

Simmel, G. (2013). Fashion. *American Journal of Sociology*, 62(6), pp. 541–558.

Sonenshein, S. (2014). How organizations foster the creative use of resources. *Academy of Management Journal*, 57(3), pp. 814–848.

Tavory, I. and Timmermans, S. (2014). *Abductive analysis. Theorizing qualitative reserach*. Chicago: The University of Chicago Press.

Thompson, J.D. (2008). *Organizations in action*, 6th ed. New Brunswick, NJ: Transaction Publishers.

Tsoukas, H. (2003). Do we really understand tacit knowledge? In: M. Easterby-Smith and M.A. Lyles, eds., *The Blackwell handbook of organizational learning and knowledge management*. Oxford: Wiley-Blackwell.

Ventresca, M.J. and Mohr, J.W. (2005). Archival research methods. In: J.A.C. Baum, ed., *The Blackwell companion to organizations*. Oxford: Blackwell Business.

# 7 Assessing values of cultural heritage and museums

## A holistic framework

*Vasilis Avdikos and Mina Dragouni*

### Introduction

Over the past decades, cultural heritage came to be an increasingly evolving concept comprised of diverse collections of movable and immovable monuments and sites, cultural practices and representations, places and landscapes, all worthy of protection (Jokilehto, 2005). In this development, museums have always played a key role, defined as the institutions that 'collect, conserve, and communicate heritage through exhibitions and display' (ICOM, 2019), with increasing emphasis on their contribution to broader societal ideals, such as cultural polyphony and social equality. This chapter explores the dynamics between the economy, society and value-making processes. By considering the numerous economic and non-economic qualities ascribed to cultural artefacts, practices, institutions, places and experiences, the chapter highlights the tensions between tradition and innovation (see also Brydges et al., this volume; Hauge and Rykkja, this volume; Merkel and Suwala, this volume; Nyfeler, this volume).

Traditionally, the conception, making and management of heritage and museums across Europe were performed largely by the state, and their future was by law safeguarded through public resources as self-evidently beneficial to society. However, recent years have marked a policy turn, expressed by a higher interest in the 'value' of heritage and museums and a shift towards an instrumental role of the sector in the economy (Pasikowska-Schnass, 2018). This shift signals the popularisation of a new rationale, whereby the sector now has a mission to serve as a pillar of socio-economic development (e.g. through urban regeneration or tourism).

Although heritage and museum management have long been imbued with the notion of value (Mason, 2002; de la Torre, 2002), the growing prevalence of neoliberal politics in public policy push for the crystallisation of a market rationale in the sector. Thus, the allocation of funds to related goods and services requires 'hard' evidence of economic and societal returns

DOI: 10.4324/9781003197065-7

(European Commission, 2014; Girard and Gravagnuolo, 2017) and policy makers are requesting research and tools that can translate both direct and broader gains into measurable objectives. This is somewhat understandable given that heritage costs are an obvious burden for governmental budgets, while benefits are less easy to capture.

Admittedly, assessing values of heritage and museums can be a particularly challenging task. Due to the inherent complexity, the scope of existing studies remains particularly narrow, as the vast majority deal mainly with economic impacts (Europa Nostra, 2015; Suwala, 2014). This downplays the plethora of other non-consumptive functions of heritage, such as sense of place, creativity, communal memory, cultural diversity and tolerance (see related discussions in Granger, this volume and Jansson and Gavanas, this volume).

As argued here, deviating from monolithic valuation approaches requires the development of more elaborate conceptual frameworks that encompass heritage and museums and their interrelationships to the social, cultural and environmental fabric of the cityscape. Yet there is a critical gap in the related literature and limited tools for the empirical assessment of the sector's value in a holistic way, which requires considering both its economic and non-economic effects (European Commission, 2018) that would in turn allow for developing better-informed policies.

In response to this, the chapter presents a novel conceptual matrix of heritage values with the view to inform empirical work on the topic. By drawing on Harvey's (2004) notions of spatiotemporality, we build a framework which locates ideas of human geography to the heritage field. In doing so, economic and non-economic value(s) associated with heritage are conceptualised together across absolute, relative and relational space. Although the matrix is applicable to various heritage domains (e.g. archaeological sites, historic monuments), we confine our analysis here to the example of museums, which represents a major component of the sector. However, our ambition is that this framework can also serve as a conceptual tool and pave the way for more holistic understandings in other domains of the creative and cultural industries (e.g. visual arts, music).

## Capturing the 'value' of cultural heritage

The term 'heritage value' is used here to describe all personal and collective benefits deriving from experiencing heritage goods, either in their physical form (such as museum collections) and/or associative meanings (Dragouni and Fouseki, 2018). Investment in the sector can impact positively on cities and their surrounding communities in multiple ways (Dümcke and Gnedovsky, 2013; Throsby, 2012). This is because it is both a capital- and labour-intensive sector, fostering direct (e.g. curation, conservation) and indirect employment in supporting and peripheral industries (e.g. tourism, leisure). At the same time, heritage and museums are closely tied to their social, cultural and environmental subsystems, playing a major role in building social capital and bonding within communities. In so doing they also serve to enhance identities, promote creativity and cultural diversity and increase aesthetic pleasure and sense of place (Murzyn-Kupisz and Działek, 2013).

Although the subject has attracted considerable attention, there is still no commonly-accepted value typology or shared understanding of the impacts assessed (Fredheim and Khalaf, 2016). Rather, the heritage literature provides highly heterogeneous distinctions of value between 'cultural/symbolic', 'historical', 'spiritual', 'aesthetic', 'iconic', 'commemorative', 'inspirational', 'recreational', 'evidential', 'scientific', 'educational', 'technological' or 'emotional', to mention but a few (McClelland et al., 2013). Parallel to these, economics-based impact studies (see for instance, Throsby, 2012; ECORYS, 2015) remain relatively partial, measuring primarily the direct contribution of the sector to local economies and some broader economic (e.g. real estate) and non-economic gains (e.g. quality of life). At the same time, there are key societal values which are largely underexplored due to conceptual complexity and methodological vagueness.

As demonstrated by existing work, impacts that translate directly into monetary figures can be captured through various indicators (e.g. level of income, employment and investment in related projects) allowing for data collection and calculations based on actual market information. For impacts that reflect non-consumptive values, some form of numerical interpretation (i.e. economic valuation) is normally pursued (Ahlfeldt, Holman and Wendland, 2012). Impact evaluation techniques that are commonly employed for expressing non-market impacts monetarily are revealed preferences techniques, such as hedonic pricing (see for example the study of Lazrak et al., 2014 on historic sites) and stated preferences methods, such as contingent valuation (see for instance the work of Lampi and Orth, 2009 on museums). Notably, most recent studies seek to link socio-cultural impacts to improved subjective well-being and quality of life of heritage users (Fujiwara, Cornwall and Dolan, 2014), an approach largely limited by considering only formal participation to heritage and thus disregarding externalities that also affect non-audiences, such as heritage wider contribution to cultural diversity and tolerance (Cervelló-Royo, Garrido-Yserte and Segura-García del Río, 2012).

Instead of using quantitative tools, another strand of the literature adopts qualitative approaches to the subject (see indicatively Lekakis, 2013; Smith, 2015; Gao, 2016; Beeksma and De Cesari, 2019; Lekakis and Dragouni, 2020). Said studies do not seek to monetise value but rather to develop in-depth understanding of heritage significance by drawing on non-statistical data collected mainly through participatory techniques. Data co-creation with user communities aligns well with current trends in the field, which advocate for a social turn towards heritage management (Council of Europe, 2000; UNESCO, 2007). Similar to social/cultural anthropology research, popular participatory tools for gaining insights into heritage values include interviews, focus groups and participant observation. A key benefit of qualitative analysis is that collected data is not considered separately from its social context and thus maintains validity and detail. However, it still remains unclear how qualitative analysis can be combined with quantitative evidence of heritage value in a constructive way.

Overall, the majority of existing economics-based studies are quite narrow in scope, capturing only a partial set of heritage values each time. As it was recently reported, the vast majority of heritage impact studies assess economic impacts exclusively or provide sectorial views instead of holistic evaluations of the societal significance of heritage (Europa Nostra, 2015). Regarding research methods, there are several quantitative and qualitative techniques, which are nonetheless fragmented. Moreover, the abundance of heritage value typologies hinders the development of a multi-dimensional conceptual framework even further. We therefore argue that maintaining dichotomies across value domains (e.g. economic, social, cultural) under various subjective taxonomies is perhaps not the way forward. Rather, approaching the question through broader analytical, interdisciplinary and holistic approaches may offer new insights that can help us deal with the inherent complexities of the problem.

## Theoretical framework; a tripartite division of space

In a seminal paper, David Harvey (2004) discusses the way he conceptualises space, building on Henry Lefebvre's distinction of space. In particular, Harvey's understanding of space consists of three main domains. The first domain is *absolute space*, a fixed universe where everything rests autonomously. We can easily recognise absolute space in positivism or as Harvey suggests in the work of Newton and Descartes; a pre-existing, static collection of events and phenomena amenable to calculation and standardised measurement.

However, as we move from the absolute to the relative domain, simplicity transforms into multiplicity. With *relative space* as the name implies, we start dealing with relationships. Objects and phenomena exist in themselves and in inter-connections. Relative space is mainly associated with Einstein and the non-Euclidean geometries. As Harvey explains, 'space is relative in a double sense: there are multiple geometries from which to choose and the spatial frame depends crucially upon what it is that is being relativised and by whom' (2004, p. 3). All forms of measurement are dependent upon the frame of reference of the observer. This illustrates that the very nature of social reality is different here, as we abandon objectivity for the sake of subjective interactivity.

As we move further to *relational space*, objects and events represent and exist only in their inter-relationships; there is no such thing as space outside of the processes that define it. Relational space is embedded or is internal to processes, which define their own spatiotemporal frameworks. At this point, we cease thinking in causal terms (causative structures, where x and y causes z) and get interested in dialectical relations—the transient aspects of society that are no longer external but become internalised, historically developing forms in motion. This ontological position is clearly inspired by the dialectics of Marx (e.g. everything is in process, value does not exist unless it is in motion).

A simple example can illustrate the above layers and assist us to better understand their meaning: in absolute space a museum's surface can be measured and mapped as well as the number of its collections and exhibited artefacts. In relative space we can compare a number of variables between different museums, or we can draw maps relating each museum's exhibition room (and their different collections) to each other or attempt to research the relation between different collections and the number of visitors (and their socio-economic backgrounds), respectively. However, in relational space the different exhibition rooms and their artefacts become meaningful when people visit or interact with the museum. Sparked by the very essence of artefacts and their symbolisms, people produce different kinds of spatiotemporal relations that develop further experiences, meanings, perceptions, memories, identities, ideologies and so on. In the relational dimension a space becomes a museum through the existence of the human factor and its cognitive and perceptual dynamics.

## From space to value: a conceptual matrix

This tripartite division of space may help us develop a new conceptualisation about how values are being revealed and evaluated in processes of cultural production and consumption. In parallel with the tripartite division of space we may think about an analogous division of cultural value, where a clear relation between space and value creates some new distinctions that could pave the way for a new holistic understanding of the concept. In order to do so, a more detailed example will help us define the interactions between space and (economic and non-economic) cultural values.

Table 7.1 presents a matrix where the absolute, relative and relational space meet the economic and non-economic values produced by the

*Table 7.1* Matrix of economic and non-economic values against a tripartite division of space

| | *Absolute* | *Relative* | *Relational* |
|---|---|---|---|
| Economic | Museums as independent and bounded entities<br>Financial analysis drawing on specific data (e.g, number of employees, ticket sales etc.) | Rankings, maps and indicators relative to the museums' contribution to the local and regional economy, ticket sales, popularity etc. Market-based evaluation techniques, stated preference methods, local macro-economic impact analyses | Economic and noneconomic values are relationally connected and cannot be separated (e.g. through experiences, perceptions and memories, in identities etc.)<br>Methodologies may include a mix of cognitive, observational, historical, ethnographic, discourse, phenomenological approaches along with quantitative market based and non-market based evaluation methods |
| Non-economic (cultural, social, environmental values) | Number of school visits, number of educational programmes, CO2 emissions | Non-consumptive values (e.g. identity building, satisfaction etc.) are monetised through quantitative methods, such as subjective well-being, or stated preferences (e.g. willingness to pay), and optimal simulation theory. Approaches may include social cost and benefit analyses, community impact analyses, comparative studies of energy performance of museums' infrastructure etc. | |

operation of a museum. In absolute space the economic values produced can be easily calculated through its balance sheets and other financial accounts, including revenues, the number of employees, visitors, ticket sales and so on. Moreover, the social, cultural and environmental footprint (non-economic values) of the museum's operation can also be measured through numerical indicators, such as the number of school visitors, cultural events, educational programmes for different age groups and the like. The museum in this example is a bounded and autonomous entity and all (absolute) values are open to calculation.

The measurement of values gets more difficult to identify and assess when we move to relative space and attempt to view and compare the production of economic and, even, cultural and social values. The museum of the previous example no longer stands alone while the production of values (or its impacts) may extend beyond its territorial boundaries. When we think of the economic values of a museum's operation in relative space, we can calculate the contribution of the museum to the local and regional and even national economy. Moreover, we can research the contribution of the museum sector in other sectors of the economy (and vice versa), using input-output tables. By extension, we can produce rankings through several indicators (most popular, most visited, most profitable museum etc). In that sense, relative values can be also illustrated through mapping techniques.

Everything can be relativised, even social and cultural values, such as identity building or cultural participation, through the use of proxies. What is of importance here is the relativisation of non-economic values through methodologies that usually attempt to monetise the outcomes of social relationships and interactions with heritage, through quantitative tools such as willingness to pay, hedonic pricing, subjective wellbeing, balance theory and optimal simulation theory. This is extremely useful for policy-makers and also for researchers as monetisation brings hard-data to the production of tangible accounts that cannot be easily controverted. However, the monetisation of cultural, social and environmental values pushes them to the sphere of economic values under a unified endeavour. In doing so, the economic space (and thinking) conquers social space, or as Lefebvre (1991) put it, the conceived space captures the lived space. This has some wider implications for the holistic understanding of heritage value as the progressive submission of cultural, social and environmental values to their economic representations, disrupts lived space (the everyday practices that produce values) and some certain qualitative aspects of it that cannot be monetised are left out. Moreover, relativism's focus on outputs and outcomes of social relations does not leave much space for the processes that define these outcomes/outputs and cannot consider the ways that non-consumptive values of heritage are carried and transformed in time-space. The participation of a teacher in a cultural event in a museum can produce a set of cultural values (output) for her, and she may transfer these values to her class, in other cultural ways and modes of contact with the pupils (processes), fostering the (co-) creation of other cultural values, and so on.

This produces a relativist trap, where the non-economic/non-consumptive values, especially those that are based on lived experiences cannot be fully addressed and explored through monetisation processes via relative economic models and quantitative methodologies.

Such a relativist trap can be avoided by moving to relational space, as relationality dictates moving from the outcomes of non-economic relationships to the processes that produce and define them. These continuous processes include actors, objects and artefacts, the built and natural environment and human or virtual networks, and so on. Identity building or co-creation practices represent socio-spatial processes that need more qualitative methodologies to unpack them and produce meaning. Participatory action research (Greenwood, Whyte and Harkavy, 1993) and multiple qualitative methods can be of value here as they can enable the researcher to explore the ways in which museums contribute to identity building and decode the relations that produce (multiple) identities over time and space. In relational space, non-economic values cannot be easily separated from the economic ones, and individual and collective perceptions of identity represent a continuous process that develops in relation to the economic, social and cultural lived space.

Even when we want to research the economic values of a museum, relational space thinking can help us unpack the ways that economic activity assimilates into the various complex social, cultural and ecological webs that surround it. Here the economic sphere should be regarded as just another sphere that needs to be analysed in relation to the social and the cultural spheres in a dialectic way. Moreover, relational space opens up the discussion of power relations and as Massey (1994) highlights, interactions in relational space should always be regarded as power relations that produce multiple power geometries. 'The means by which people are 'placed' within given sets of relations can either strengthen or weaken their ability to exercise some degree of control over those very relations' (1994, pp. 28–30). Thus, the ways in which certain communities think about the economic contribution of a museum to their local economy is constantly in relation to the overall position of the museum to their perceptions. Indeed, a museum can play the role of a cultural flagship in a city, which can be associated with vast private investments, gentrification and processes of social exclusion. Thus, the citizens may perceive the museum's economic contribution in relation to the socio-economic and political power geometries entailed in its development, and in relation to the ways in which museums are embedded in local and supra-local economic networks.

Furthermore, a relational perspective can help when we need to explore the impacts on communities that engage with heritage in abstraction. As residents of Athens, we might have visited the Acropolis Hill once or twice throughout our lifetime, but in reality, we encounter it multiple times every day. We also constantly come across depictions of the Parthenon or even minimalist symbols of antiquity, such as a Doric column. How do these processes of symbolic reception play out for an Athenian, a Greek, a Greek of the Diaspora, a refugee, an immigrant, a European tourist, a visitor from the US or China? Where do these narratives come from, how have been shaped historically, how do people use them to interpret the present, how do they transform these narratives and how do the narratives transform them? Understanding these processes of interactivity holds great potential value for cultural policy across all levels. For instance, a Doric column, as a globally recognised motif, can be used as a tool in an AI video game that aims to ease refugees' social transition from the Eastern Mediterranean to the host countries of the Balkans. Relational thinking and methodologies can provide a useful conceptual terrain for engaging with complex social and cultural relations and their fluid impacts that relative space and thinking cannot unveil on their own.

## Conclusion

Capturing the value of cultural heritage becomes increasingly important for related cultural and development policies so that decisions of resource allocation can be well-informed and accountable. However, current work on the subject is still limited by sectorial views and mono-dimensional interpretations of value creation and reproduction processes. Accounting for the economic and broader social impacts (e.g. cultural, environmental) of heritage by existing studies is extremely limited and remains confined mostly to impacts realised in absolute and relative space.

The tripartite division of cultural values that the paper puts forward aims to contribute to the development of a more holistic framework for the analysis of economic and non-economic values in the field of museums and cultural heritage. However, this framework can also be used beyond cultural heritage; in other and broader spheres of cultural production and consumption, where the epicentre is cultural value production and valuation. What is of importance here is that geography and geographical thinking can enrich our understandings of how values are produced; as a synthesis of multiple dynamics that operate in different levels (the social, the economic, the cultural and so on) and different spatial levels (from the local to the global). Thus, we need to synthesise all levels in order to arrive at a holistic framework that can provide answers to researchers and policy makers.

We believe that relationality and relational space can broaden enquiry upon cultural value production (see also Comunian et al., this volume; Hauge and Rykkja, this volume; Jansson and Gavanas, this volume).

However, this new framework needs to be tested empirically, in order to develop methodological devices where the three spatial levels (absolute, relative, relational) can be synthesised in case studies. Their synthesis can be done in a nested way, where the absolute is included in the relative, and the relative is embraced by relational space. A potential entry point to such empirical exploration might be to use information from absolute and relative accounts in order to inform relational enquiries about value production.

For instance, relational thinking for researching co-creation processes in museums (e.g. individual and community interpretations of exhibits), should always consider the economic, cultural and social background of those that usually visit the museum. Here, relativist accounts and maps of museum visitors can provide useful basic information before researching the ways in which co-creation processes can be developed. Similarly, policies that aim to foster social inclusion through cultural events and heritage engagement, should also search for the human geography of those excluded and the subsequent power geometries, before planning relevant strategies. Beyond thinking of these three divisions in a hierarchical way, we believe that a dialectical approach can produce more fruitful results, or at least discussions upon cultural value production. Dialectics across the three divisions and economic/non-economic values can arise when power relations are found to play a distinctive role, not only in the social and cultural fabric of relational spaces through the multiple power geometries, but also in the ways that economic space fosters or hinders specific cultural and social values.

This chapter introduces the ways (relational) geography can enrich the discussion of economic and non-economic values in the field of cultural heritage but more work needs to be done in order to elaborate on this framework conceptually and methodologically. Admittedly, cultural heritage and its tangible and intangible (traditions, knowledge, skills, practices) properties plays a pivotal role in value-creating processes and the production of idiosyncratic culture representations of places and communities. Increasing our theoretical understanding of these processes would be valuable for reconciling the tensions between innovation, development and the past; drawing instead on their relational complementarities to produce policy instruments to support and sustain creative economies.

## References

Ahlfeldt, G.M., Holman, N. and Wendland, N. (2012). *An assessment of the effects of conservation areas on value*. London: English Heritage.

Beeksma, A. and De Cesari, C. (2019). Participatory heritage in a gentrifying neighbourhood: Amsterdam's Van Eesteren museum as affective space of negotiations. *International Journal of Heritage Studies*, 25(9), pp. 974–991.

Cervelló-Royo, R., Garrido-Yserte, R. and Segura-García del Río, B. (2012). An urban regeneration model in heritage areas in search of sustainable urban development and internal cohesion. *Journal of Cultural Heritage Management and Sustainable Development*, 2(1), pp. 44–61.

Council of Europe. (2000). *Landscape convention*. Available at: www.coe.int/en/web/landscape/.

de la Torre, M., ed. (2002). *Assessing the values of cultural heritage. Research report*. Los Angeles: The Getty Conservation Institute.

Dragouni, M. and Fouseki, K. (2018). Drivers of community participation in heritage tourism planning: An empirical investigation. *Journal of Heritage Tourism*, 13(3), pp. 237–256.

Dümcke, C. and Gnedovsky, M. (2013). The social and economic value of cultural heritage: Literature review. *EENC Paper*, pp. 1–114.

ECORYS. (2015). *Survey of listed buildings owners: A final report submitted to historic England*. Available at: https://historicengland.org.uk/content/heritage-counts/pub/2015/listed-building-owners-survey-2015-pdf/

Europa Nostra. (2015). *Cultural heritage counts for Europe*. Krakow: International Cultural Centre.

European Commission. (2014). *Towards an integrated approach to cultural heritage for Europe*. Brussels: European Commission, July.

European Commission. (2018). *Innovation & cultural heritage*. Royal Museum of Arts and History, Brussels—Conference Report.

Fredheim, L.H. and Khalaf, M. (2016). The significance of values: Heritage value typologies re-examined. *International Journal of Heritage Studies*, 22(6), pp. 466–481.

Fujiwara, D., Cornwall, T. and Dolan, P. (2014). *Heritage and wellbeing*. Swindon: English Heritage.

Gao, Q. (2016). Social values and archaeological heritage: An ethnographic study of the Daming Palace archaeological site (China). *European Journal of Post-Classical Archaeologies*, 6, pp. 213–234.

Girard, L.G. and Gravagnuolo, A. (2017). Circular economy and cultural heritage/landscape regeneration. Circular business, financing and governance models for a competitive Europe. *BDC. Bollettino Del Centro Calza Bini*, 17(1), pp. 35–52.

Greenwood, D.J., Whyte, W.F. and Harkavy, I. (1993). Participatory action research as a process and as a goal. *Human Relations*, 46(2), pp. 175–192.

Harvey, D. (2004). *Space as a keyword*. Paper for Marx and Philosophy Conference, Institute of Education, London, May 29. Available at: frontdeskapparatus.com/files/harvey2004.pdf.

ICOM. (2019). *Creating the new museum definition*, April 1. Available at: https://icom.museum/en/news/the-museum-definition-the-backbone-of-icom/.

Jokilehto, J. (2005). *The definition of cultural heritage: References to documents in history*. ICCROM Working Group 'Heritage and Society'. Available at: http://cif.icomos.org/pdf_docs/Documents%20on%20line/Heritage%20definitions.pdf

Lampi, E. and Orth, M. (2009). Who visits the museums? A comparison between stated preferences and observed effects of entrance fees. *Kyklos*, 62(1), pp. 85–102.

Lazrak, F., Nijkamp, P., Rietveld, P. and Rouwendal, J. (2014). The market value of cultural heritage in urban areas: An application of spatial hedonic pricing. *Journal of Geographical Systems*, 16(1), pp. 89–114.

Lefebvre, H. and Nicholson-Smith, D. (1991). *The production of space* (Vol. 142). Oxford: Wiley-Blackwell.

Lekakis, S. (2013). Distancing and rapproching: Local communities and monuments in the Aegean Sea—a case study from the island of Naxos. *Conservation and Management of Archaeological Sites*, 15(1), pp. 76–93.

Lekakis, S. and Dragouni, M. (2020). Heritage in the making: Rural heritage and its mnemeiosis at Naxos island, Greece. *Journal of Rural Studies*, 77, pp. 84–92.

Mason, R. (2002). Assessing values in conservation planning: Methodological issues and choices. In: M. de la Torre, ed., *Assessing the values of cultural heritage: Research report*. Los Angeles: Getty Conservation Institute, pp. 5–30.

Massey, D. (1994). *Space, place and gender*. Cambridge: Polity Press.

McClelland, A., Peel, D., Hayes, C.M. and Montgomery, I. (2013). A values-based approach to heritage planning: Raising awareness of the dark side of destruction and conservation. *Town Planning Review*, 84(5), pp. 583–604.

Murzyn-Kupisz, M. and Działek, J. (2013). Cultural heritage in building and enhancing social capital. *Journal of Cultural Heritage Management and Sustainable Development*, 3(1), pp. 35–54.

Pasikowska-Schnass, M. (2018). *Cultural heritage in EU policies. European parliamentary research service*. Available at: www.europarl.europa.eu/RegData/etudes/BRIE/2018/621876/EPRS_BRI(2018)621876_EN.pdf.

Smith, L. (2015). Theorizing museum and heritage visiting. In: A. Witcomb, and K. Message, eds., *The international handbooks of museum studies*, New Jersey: John Wiley & Sons, pp. 459–484.

Suwala, L. (2014). *Kultur, Kreativität und Raum—ein wirtschaftsgeographischer Beitrag am Beispiel des kulturellen Kreativitätsprozesses*. Wiesbaden: Springer.

Throsby, D. (2012). *Investment in urban heritage. Urban development & local government*. Washington, DC: The World Bank.

UNESCO. (2007). *Convention on the protection and promotion of the diversity of cultural expressions*. Available at: https://unesdoc.unesco.org/ark:/48223/pf0000142919.

# 8 Spatial processes of translation and how coworking diffused from urban to rural environments

## The case of Cowocat in Catalonia, Spain

*Ignasi Capdevila*

### Introduction

The organisation of work is changing. In intensive knowledge-based sectors, and especially in the creative industries, an increasing need for flexibility and new sources of creativity have led.to an extensive projectification of work, and flexible arrangements combining diverse professional specialisations. Creative workers (either freelancers, entrepreneurs, or teleworkers) have experienced the need to find new working environments that inspire their creativity and, at the same time, facilitate socialisation and collaboration with peers. In the last ten years, there has been an increasing number of places that host different forms of collaborative work, like coworking spaces, that initially emerged in cities, and represented new places of coordination within urban contexts, displaying coordinating functions among urban actors.

Nevertheless, despite of the fact the development of innovations and creativity has been referred mainly as an urban phenomenon (Florida, 2002), recent accounts underline the importance of peripheries in innovation processes (Shearmur and Doloreux, 2016). In recent years, the expansion of high-speed internet together with the progressive acceptance of teleworking, have convinced many professionals to locate outside central areas, thus contributing to the launch of collaborative spaces in rural and peripheral areas.

In this context, this chapter analyses the translation process of coworking practices from an urban setting, Barcelona's metropolitan area, into rural areas in Catalonia. In doing so, the chapter explores the tension between individual and collaborative cultural labour (tension 1) and in particular, the role of new spaces of creative work play in rural environments. Indeed, a key contribution of the chapter is the exploration of coworking spaces outside of urban centres, a theme which also connects to tension

DOI: 10.4324/9781003197065-8

3 (the tension between isolated and interconnected spaces of creativity). The results suggest that through a process of reinterpretation, the different aspects constituting the understanding of coworking (materiality, practices, and values) were able to be disembedded from its original urban context to be re-embedded in different rural environments.

In order to understand the transformation of the practice of coworking, the theoretical background for this research is premised on the concept of translation (Latour, 1986), denoting the transfer and adaptation of an idea into a different form via a process involving a chain of actors and the transformation of an idea into a practice (Czarniawska and Joerges, 1996; Sahlin-Andersson and Sevón, 2003).

When analysing the evolution of the translation of coworking from urban to rural areas, coworking can be considered at different levels: coworking as sharing a physical space and working tools (materiality) (Cnossen and Bencherki, 2019), coworking as a working practice based on collaboration among peers (practice) (Garrett, Spreitzer and Bacevice, 2017; Jakonen et al., 2017), and coworking as a concept related to values and principles linked to the gift economy, the sharing economy, and the hacker principles (values). The case of the Cowocat Rural shows how the process of diffusion of coworking to rural areas was based on a progressive understanding of the coworking through a collective process of translation. Further explorations of these themes in this volume include Jansson and Gavanas who explore how subcultural expressions flow from global urban cultural nodes to the periphery, and Merkel and Suwala who look within coworking spaces to explore how intermediaries construct collaborative work atmospheres.

## Literature review

### *Translating knowledge in spatial contexts*

A dominant perspective in the literature on the diffusion of innovation (Hägerstrand, 1966; Rogers, 2010) conceives adopters as rational actors that, through a series of information cascades, scan and select new knowledge. An innovation is finally adopted once its benefits are confirmed. Rational and institutional accounts either emphasise the broad access to general information or the conformity pressures to imitation, but both perspectives focus on the conditions for diffusion.

This dominant approach consists of arguing that knowledge needs to be codified in order to be able to be transferred to distant contexts. However, the transmission of tacit knowledge—knowledge related to practice and hard to be codified—needs geographical proximity. From a territorial perspective, knowledge shared by co-located actors allows the development of localised learning capacities (Maskell and Malmberg, 1999). Through being embedded in the local "buzz" (Bathelt, Malmberg and Maskell, 2004; Storper and Venables, 2004) actors participate automatically in the sharing of ideas.

However geographical proximity is not enough, as territories require to be open to the influence of external knowledge in order to be competitive. Actors need to be able to create the required "global pipelines" (Maskell, Bathelt and Malmberg, 2006) with external actors in order to capture new knowledge by temporally re-locating, as is the case, for instance, with professional trade fairs, thus creating temporary co-location (Bathelt and Schuldt, 2008; Li, 2014).

Studies on proximities have clarified some misconceptions. Firstly, that the dichotomy between tacit/codified knowledge is not directly linked to the geographical proximity. Codified knowledge, as well as tacit, is highly contextualised (Howells, 2012) and its absorption also requires socialisation and different types of proximity (not necessarily geographical) (Rutten, 2017). This implies that knowledge creation and learning happen in social contexts, such as networks (Bathelt, Malmberg and Maskell, 2004), communities of practice (Wenger, 1998), or epistemic communities (Cohendet et al., 2014; Håkanson, 2005).

Nevertheless, despite the social aspect of its creation and diffusion, previous accounts have considered knowledge as a black box that remains unchanged in the diffusion process. This is in part due to the fact that the literature has traditionally taken a rationalistic perspective considering that knowledge is "created", "transferred", or "used" (Hautala and Jauhiainen, 2014) considering it as static and largely ignoring its social and interpretative aspect. Knowledge is a process connected to different dimensions of time (Hautala and Jauhiainen, 2014; Ibert, Hautala and Jauhiainen, 2015), in the same way that it is related to the spatial context of creation. The act of knowing, as an action and process, has to be considered in its social and time-spatial context (Ibert, 2007).

### *The sociality of spatial knowledge dynamics*

Previous research on knowledge flows has focused on the diffusion process at the scale of the socio-spatial contexts of interaction (Rutten, 2017). However, little attention has been paid to the adaptation and the changes of knowledge while it is diffused to different contexts. Here, research in management and sociology has showed that the adaptation process may involve a change in the practice during the adaptation phase and once adopted (Ansari, Fiss and Zajac, 2010), as well as different versions of a same practice in different contexts.

Practices are likely to evolve during the implementation process, requiring an adaptation to another context. Building on translation theories (Callon, 1986; Latour, 1986), researchers in the tradition of Scandinavian institutionalism (Czarniawska and Joerges, 1996; Sahlin and Wedlin, 2008) have referred to translation as a process of disembedding and re-embedding of an idea in a different time and space. Similarly, in the study of knowledge-based relationships between actors in transnational contexts, Bathelt, Cantwell and Mudambi (2018) suggest that complex translation processes can be decomposed in three phases: 1) connecting, 2) sense-making, and 3) integrating. We integrate these process model in our analysis as it reflects both the cognitive evolution of involved agents and their efforts to progressively adapt existing knowledge into a new setting.

It is in this context this chapter examines how knowledge is translated in different geographical contexts. The chapter analyses practices of coworking in Catalonia which, after an intense development in Barcelona's metropolitan area, were progressively adopted and adapted to Catalan rural and peripheral areas. Through the process translation, the understanding of coworking changed through a number of phases of development, resulting in an interpretation of the coworking practices adapted to the characteristics of rural environments.

## *The elements of coworking: materiality, practice, and values*

Over the last ten years, coworking has emerged as a new away if working, defined by values as autonomy in work, empowerment, collaboration, and sense of community (Garrett, Spreitzer and Bacevice, 2017; Merkel, 2015, 2019; Schmidt, Brinks and Brinkhoff, 2014; Spinuzzi et al., 2019; Vidaillet and Bousalham, 2018). Coworking as a practice has been translated to an explosion of these spaces all over the world.

Coworking spaces are collaborative spaces for working and socialising where, by paying a monthly fee, members have access to a desk and shared resources, but also to an entrepreneurial ecosystem and a community open to knowledge sharing and collaboration (Gandini, 2015). It is generally accepted that the first coworking space was founded in 2005 in San Francisco (Vidaillet and Bousalham, 2018), even if Barcelona-based Kubik argues it opened as a coworking space ten years before (www.kubikbcn.com). The number of spaces increased globally from 600 in 2010 to 15,500 at the end of 2017, accounting for 1.74 million workers, four-fifths of whom are based in Europe or North America (Deskmag, 2018).

The conceptual decomposition of coworking (into materiality, practice, and values) helps to explain the phenomenon at different levels and to understand how the different aspects hold varying degrees of prominence in the different stages of its diffusion.

### *Materiality*

From Oldenburg's seminal work (2002), the concept of third place highlights the importance of the physical space (between home and the workplace) that facilitates socialisation, nurturing face-to-face interactions and fostering the emergence of unexpected and serendipitous encounters (Jakonen et al., 2017). Conceptualising coworking as "working alone together" (Spinuzzi, 2012) resonates with the idea that coworking is mainly about physical co-location. From this perspective, coworking spaces are similar to shared offices, centred on the sharing of material infrastructure (from printers to coffee machines) to reduce costs or have access to central locations.

### *Practice*

Practitioners and academics alike acknowledge that coworking spaces are more than merely shared offices (Garrett, Spreitzer and Bacevice, 2017; Jakonen et al., 2017; Spinuzzi et al., 2019). Coworking is a way of working that emphasises innovation through collaboration. From this perspective, coworking refers to a practice (coworking as an activity) rather than a space. Community managers play a crucial role, contributing to the creation of collaborative relationships and professional linkages among coworkers (Garrett, Spreitzer and Bacevice, 2017). Coworking spaces are consequently platforms where individuals with different profiles gather and interact (Vidaillet and Bousalham, 2018).

### *Values*

Coworking as a global movement, is linked to other phenomena like the sharing economy, the gig economy, or the gift economy (Bouncken and Reuschl, 2018). The movement can be defined as "a new economic engine composed of collaboration and community" (Coworking.com, n.d.). The literature on coworking has been defined around the values of collaboration, community building, knowledge sharing, and peer support (Capdevila, 2015; Gandini, 2015; Vidaillet and Bousalham, 2018). Coworking is described as a new way of working, adapted to the current context of increasing independent and intermittent work. Far from being the panacea in the organisation of flexible work relations, coworking also reflects the precarity of independent workers and the dark side of Florida's creative class (Vidaillet and Bousalham, 2018). Gandini suggests that coworking could give rise to a new class consciousness accompanied by "the rise of these atomized entrepreneurial subjects of neoliberalism" (2015, p. 202).

## Methodology

This chapter draws upon a five-year study of coworking in Barcelona and Catalonia. Data was collected in three phases. The first phase, from 2013 to 2015, consisted in a research project on the coworking spaces in Barcelona including 28 interviews with managers and coworkers. Data triangulation was done with other nine interviews to academics, policy makers, and specialists highly acknowledgeable of the Barcelona innovation dynamics.

In a second phase, from 2015 to 2017, a new round of interviews was done to 27 managers and coworkers of Catalan rural spaces, as well as representatives of the Catalan government, regional coordinators, and managers of Cowocat Rural.

In a third phase, five follow-up interviews were done with the coordinators of the Cowocat Rural, representatives and some of the spaces' managers, with the goal to have a longitudinal understanding of the evolution of the network of rural spaces.

Finally, archival data (like materials for presentations and for managers' training, established criteria for selection of coordinators, guidelines to managers, etc.) and online content of the different spaces was used to understand the different stages in the development of coworking in rural areas.

To analyse the data from a spatiotemporal perspective a number milestones were identified, including the progressive opening of rural spaces, along with the evolution in the discourses around coworking from policy makers and managers of Cowocat Rural. That first analysis sought to identify the progressive conceptualisation and implementation of rural coworking in Catalonia. Second, the analysis focused on gaining a deeper understanding of the role of the different actors (municipal, regional, and governmental policy makers, space managers, and coordinators of the Cowocat Rural network) in the translation of urban coworking in rural areas. Third, taking as a framework of analysis the different aspects that constitute the concept of coworking (materiality, practice, and values) and the theoretical process model (Bathelt, Cantwell and Mudambi, 2018), the different stages in the translation process (connecting, sensemaking, and integrating) were characterised to understand how the knowledge associated to urban coworking was transformed to be adapted to different rural contexts.

## The diffusion of coworking to rural areas: the case of Cowocat Rural

Over the last decade, Barcelona has become one of the most important European hubs for coworking (Coll-Martínez and Méndez-Ortega, 2020; Institut Cerdà/AMB, 2019). Two decades later, there were over one hundred spaces using the term coworking to define themselves (Capdevila, 2015; Institut Cerdà/AMB, 2019), even if the number is highly variable, and likely to be strongly affected by the effects of the COVID-19 crisis. From the 2010s, public bodies (notably the Catalan Government) started to be interested in coworking, considering it a way to boost entrepreneurship and to help unemployed people to join professional working settings. The Catalan Government promoted the creation of Cowocat, to capitalise best practices and to create synergies among the spaces. This network of coworking spaces facilitated the diffusion of the practice of coworking in Catalonia, raising awareness and interest among town councils from rural areas attracted by the idea that coworking could contribute to retain talent and develop the local economy.

In 2011, in Riba-roja d'Ebre (a small village 200 km south from Barcelona), a group of young councillors had started to invest in a local network of optical fibre while considering the possibility of rehabilitating the former public library into a coworking space. They got in contact with technicians of the local consortium of socioeconomic initiatives "Consorci Intercomarcal d'Iniciatives Socioeconòmiques" (CIS) to assist them in determining the technical aspects of such a space. By that time, the manager and one technician of CIS were already interested in the concept of coworking gaining popularity in Barcelona. The CIS team studied the urban coworking phenomenon to adapt it to rural areas. In 2012, the coworking space (called Zona Líquida) was launched through the development of the "Consell Comarcal" (County Council—CC). The goal of the pilot project was to revitalise the territory and to strengthen the relationship among local entrepreneurs.

The project, which started as a public space open to local entrepreneurs, was often empty, due to a lack of targeted communication and a certain difficulty to identify potential users. In their beginnings, the experience did not reach the expected results. To try and engage local entrepreneurs, the promoters selected five young rural entrepreneurs and freelancers (according to their professional profile) and sent them on a trip to Barcelona to visit five different coworking spaces. They came back being convinced of the coworking "philosophy" and the importance of community management activities that had to be organised by dedicated persons. Following their recommendation, Zona Líquida started to open on a daily basis, and hired a community manager, responsible for the search for new members, and their integration by facilitating networking among members. After being almost empty for the first two years, professionals from the village and surroundings progressively started to use the space.

Then, in 2014, the CIS and CC thought that the initiative could be replicated in other Catalan rural areas and started a project open to other partners with the aim of retaining talent, attracting population and helping local entrepreneurs. That year, the project Cowocat Rural was launched, with the financial contribution of the EU funds for rural development. The project started as a network of rural coworking spaces affiliated to a larger network (Cowocat, which included mainly Barcelona-based coworking spaces). Cowocat Rural is a non-for-profit initiative with support from public institutions, notably the Catalan government, and currently consists of fifteen rural spaces.

The promoters of Cowocat Rural encouraged coworking by offering support to village councils looking to open a space. One of their main challenges has been to explain that the success of coworking depends on the pre-existence of a community. The characteristics of rural areas, with low population densities and high transport times, make the frequent co-location of members in the same space more difficult than in the urban environment. That is why the spread of coworking is not only done by opening new physical spaces, but also, and more importantly, by previously analysing the territory, its economic and industrial trajectory and its current needs and by developing the local community. In particular, the identification of existing professional communities was considered as an important precondition before taking the decision of opening a space.

A commonly identified mistake in rural coworking has been to rush into the launch of a space without considering the current needs or the potential users. Given the lack of knowledge about the values and principles of coworking, promoters engaged in the "evangelisation", both at the level of the citizenship and at the institutional level, by providing examples of real successful cases.

The project has since been extended with several side projects. For instance, an initiative called "Rural & Go" proposes temporary gatherings of one to two weeks to experience co-living in rural environments to urban coworkers. The promoters also launched a "coworking lab" in a university, where researchers analyse the collaboration dynamics in real conditions. Also, there is an initiative to develop an online platform to facilitate the interaction and collaboration among rural coworkers located in different spaces in Catalonia.

## Discussion

Based on the three stages of translation processes (Bathelt, Cantwell and Mudambi, 2018), Table 8.1 shows how the process of diffusion of coworking to rural areas was based on a progressive understanding of coworking, parallel to a reinterpretation of each characteristic to adapt it to the rural reality. In the first stage, the focus was on materiality. In that phase, a simplistic view of coworking as physical spaces allowed a rapid diffusion of the concept that led the first village councils to be attracted to it.

*Table 8.1* The Cowocat Rural case. The stages of the diffusion of coworking from urban to rural areas

| *Phase* | *Focus* | *Activities* |
|---|---|---|
| Connecting (2011–2012) | Materiality (coworking as sharing a space and assets) Focus on the space and the technologic infrastructure | • Village councillors decide to transform the library into a coworking space (2011).<br>• Local development bodies (CC & CIS) identify coworking as a way encourage professionals from urban areas to return to their villages.<br>• First Catalan rural coworking space opens (2012). |
| Sense making (2012–2014) | Practice (coworking as a new way of working based on colocation, collaborating and sharing knowledge). Focus on community building and complementarity of knowledge bases. | • Rural entrepreneurs visit Barcelona coworking spaces.<br>• Promoters understand the importance of community managers.<br>• Organisation of meetings to diffuse the practice of coworking. |
| Integrating (from 2014) | Values (coworking as a movement based on sharing and collaborating). Focus on the diffusion of common values and principles. | • Confidence that the initiative can be extrapolated to other rural areas.<br>• Creation of Cowocat Rural network.<br>• Presentation of the network in regional and national conferences.<br>• Launch of side projects (i.e. webpage with coworkers' profiles, "Rural & Go" project). |

Simply packaged ideas can travel fast in a context of an intense buzz (Rutten, 2017), however, once adopted, need to be reframed in the new context.

That is the reason why, after the connecting phase, the rural adopters had to engage in a sense-making stage, to put the concept into practice. To do so, they had to create linkages (Bathelt, Malmberg and Maskell, 2004) with distant actors to learn from them. By engaging in active conversations (Rutten, 2017), rural entrepreneurs could learn practices from their urban peers.

As the practice involved a great deal of tacit knowledge embedded in the everyday informal interaction and collaboration among peers, it required temporary co-location (Maskell, Bathelt and Malmberg, 2006) of rural coworkers discovering urban spaces, and rural promoters visiting spaces, participating in conferences and meeting urban space managers. The process of sense-making required a swift of focus, from materiality to practice, from an absorption of codified knowledge, to internalise it as tacit.

The result was an adaptation of the urban practice to the rural. For instance, rural coworking can hardly be specialised (due to population density) but, nevertheless, rural adopters could extract a high value on their diversity (Meili and Shearmur, 2019). Community management also differed from urban to rural. Whereas in urban areas, managers are confronted with a high rotation of members, rural managers are dealing with rather stable small core communities with frequent face-to-face interaction, in combination with a larger local community characterised by weak ties and punctual co-location (mainly done during events and trainings).

The third phase of the translation process consisted of the integration of new knowledge on rural coworking to the ongoing global conversation (Rutten, 2017) around coworking. At a local level, the rural initiative joined a larger network of spaces (Cowocat). On the national and international levels, the promoters diffused their success cases' experiences and lessons learned to policy makers and village councils attracted by the idea of using coworking as a way to improve their entrepreneurial and economic environment. The re-interpretation of the concept of coworking into rural areas was done by the active and purposeful action of "translators", individuals like Cowocat Rural coordinators and space managers. They further contribute to the diffusion of the values of coworking by diffusing its rural interpretation, thus reinforcing the coworking movement, beyond its urban model.

## Conclusion

Until recently, research on the geographies of creativity and innovation have focused on urban and industrial regions, assuming that there was little innovation in rural and peripheral environments. Indeed, this has been a key theme explored during the European Colloquiums on Culture, Creativity, and Economy (CCE) events more broadly (see the introduction in this volume for more details), and in other contributions to this volume, including Jansson and Gavanas. In recent years, the development of the digital infrastructures in rural areas has allowed the coordination of work at a distance and the relocation of workers out of the urban regions. This movement has especially impacted the creative industries and other industries characterised by the ad-hoc coordination of highly mobile, highly skilled professionals. Similar to urban areas where a project-based organisation of work has facilitated the emergence of coworking spaces, rural areas have also seen the co-location of independent workers, freelancers, and entrepreneurs to socialise, collaborate, share knowledge, and resources.

Nevertheless, as the results presented in this chapter show, the diffusion of the practice of coworking from urban to rural areas has not been a replication but an adaptation to a new context. In doing so, this chapter has contributed to the tension between individual and collaborative creative practice) and the tension between isolated and interconnected spaces of creativity through not only investigating new spaces of work but also moving beyond the study of urban centres (see also Granger, this volume, and Comunian et al., this volume). The results contribute to explain how translation of a practice takes place from a socio-spatial perspective and how the different stages of the translation involve different conceptualisations of the practice. This research also underlines the importance of the role of translators, the agents that actively engage in the translation, guiding the involved agents through the process. Finally, it also provides some insights about the spatial transformation of work practices in the creative and knowledge-based industries, reinforcing the idea that creative and innovative activities do not exclusively take place in urban areas. The results also show knowledge generated at the core geographies might diffuse to the periphery and provides some insights about how policy making and institutional support can contribute to develop spaces for collective innovation in rural areas.

## References

Ansari, S.M., Fiss, P.C. and Zajac, E.J. (2010). Made to fit: How practices vary as they diffuse. *Academy of Management Review*, 35(1), pp. 67–92.

Bathelt, H., Cantwell, J.A. and Mudambi, R. (2018). Overcoming frictions in transnational knowledge flows: Challenges of connecting, sense-making and integrating. *Journal of Economic Geography*, 18(5), pp. 1001–1022.

Bathelt, H., Malmberg, A. and Maskell, P. (2004). Clusters and knowledge: Local buzz, global pipelines and the process of knowledge creation. *Progress in Human Geography*, 28(1), pp. 31–56.

Bathelt, H. and Schuldt, N. (2008). Between luminaires and meat grinders: International trade fairs as temporary clusters. *Regional Studies*, 42(6), pp. 853–868.

Bouncken, R.B. and Reuschl, A. (2018). Coworking-spaces: How a phenomenon of the sharing economy builds a novel trend for the workplace and for entrepreneurship. *Review of Managerial Science*, 12(1), pp. 317–334.

Callon, M. (1986). Some elements of a sociology of translation: Domestication of the scallops and the fishermen of St. Brieuc Bay BT—power, action and belief. A new sociology of knowledge? *Power, Action and Belief. A New Sociology of Knowledge? D*, pp. 196–233.

Capdevila, I. (2015). Co-working spaces and the localised dynamics of innovation in Barcelona. *International Journal of Innovation Management*, 19(3), p. 1540004.

Cnossen, B. and Bencherki, N. (2019). The role of space in the emergence and endurance of organizing: How independent workers and material assemblages constitute organization. *Human Relations*, 72(6), pp. 1057–1080.

Cohendet, P., Grandadam, D., Simon, L. and Capdevila, I. (2014). Epistemic communities, localization and the dynamics of knowledge creation. *Journal of Economic Geography*, 14(5), pp. 1–26.

Coll-Martínez, E. and Méndez-Ortega, C. (2020). Agglomeration and coagglomeration of co-working spaces and creative industries in the city. *European Planning Studies*, pp. 1–22.

Coworking.com. (n.d.). *Coworking.com wiki*. Available at: http://wiki.coworking.com/w/page/16583831/FrontPage#WhatisCoworking.

Czarniawska, B. and Joerges, B. (1996). Travel of ideas. In: B. Czarniawska and G. Sevo, eds., *Translating organizational change*. Gothenburg: University-School of Economics and Commercial Law/Gothenburg Research Institute, pp. 13–47.

Deskmag. (2018). *Ultimate cowork data report*. Available at: www.deskmag.com/en/the-state-of-coworking-spaces-in-2018-market-research-development-survey.

Florida, R. (2002). Bohemia and economic geography. *Journal of Economic Geography*, 2(1), pp. 55–71.

Gandini, A. (2015). The rise of coworking spaces: A literature review. *Ephemera: Theory & Politics in Organization*, 15(1), pp. 193–205.

Garrett, L.E., Spreitzer, G.M. and Bacevice, P.A. (2017). Co-constructing a sense of community at work: The emergence of community in coworking spaces. *Organization Studies*, 38(6), pp. 821–842.

Hägerstrand, T. (1966). Aspects of the spatial structure of social communication and the diffusion of information. *Papers in Regional Science*, 16(1), pp. 27–42.

Håkanson, L. (2005). Epistemic communities and cluster dynamics: On the role of knowledge in industrial districts. *Industry and Innovation*, 12(4), pp. 433–463.

Hautala, J. and Jauhiainen, J.S. (2014). Spatio-temporal processes of knowledge creation. *Research Policy*, 43(4), pp. 655–668.

Howells, J. (2012). The geography of knowledge: Never so close but never so far apart. *Journal of Economic Geography*, 12(5), pp. 1003–1020.

Ibert, O. (2007). Towards a Geography of Knowledge Creation: The Ambivalences between 'knowledge as an object' and 'knowing in practice'. *Regional Studies*, 41(1), pp. 103–114.

Ibert, O., Hautala, J. and Jauhiainen, J.S. (2015). From cluster to process: New economic geographic perspectives on practices of knowledge creation. *Geoforum*, 65, pp. 323–327.

Institut Cerdà/AMB. (2019). *Els espais de coworking a les ciutats ". Informe d'aprofundiment de l'economia metropolitana* (No. 17, p. 19). Institut Cerdà/Àrea de Desenvolupament Social i Econòmic de l'AMB. Available at: amb.cat/es/web/desenvolupament-socioeconomic/actualitat/publicacions/detall/-/publicacio/els-espais-de-coworking-a-les-ciutats/7982091/11708.

Jakonen, M., Kivinen, N., Salovaara, P. and Hirkman, P. (2017). Towards an economy of encounters? A critical study of affectual assemblages in coworking. *Scandinavian Journal of Management*, 33(4), pp. 235–242.

Latour, B. (1986). The powers of association. *The Sociological Review*, 32(1), pp. 264–280.

Li, P.-F. (2014). Global temporary networks of clusters: Structures and dynamics of trade fairs in Asian economies. *Journal of Economic Geography*, 14(5), pp. 995–1021.

Maskell, P., Bathelt, H. and Malmberg, A. (2006). Building global knowledge pipelines: The role of temporary clusters. *European Planning Studies*, 14(8), pp. 997–1013.

Maskell, P. and Malmberg, A. (1999). The competitiveness of firms and regions: 'Ubiquitification' and the importance of localized learning. *European Urban and Regional Studies*, 6(1), pp. 9–25.

Meili, R. and Shearmur, R. (2019). Diverse diversities—Open innovation in small towns and rural areas. *Growth and Change*, 50(2), pp. 492–514.

Merkel, J. (2015). Coworking in the city. *Ephemera: Theory & Politics in Organization*, 15(1).

Merkel, J. (2019). 'Freelance isn't free.' Coworking as a critical urban practice to cope with informality in creative labour markets. *Urban Studies*, 56(3), pp. 526–547.

Oldenburg, R. (2002). *Celebrating the third place: Inspiring stories about the 'great good places' at the heart of our communities*. Boston: Da Capo Press.

Rogers, E.M. (2010). *Diffusion of innovations*. New York: Simon and Schuster.

Rutten, R. (2017). Beyond proximities: The socio-spatial dynamics of knowledge creation. *Progress in Human Geography*, 41(2), pp. 159–177.

Sahlin, K. and Wedlin, L. (2008). Circulating ideas: Imitation, translation and editing. In: R. Greenwood, C. Oliver, K. Sahlin, and R. Suddaby, eds., *The SAGE handbook of organizational institutionalism*. London: Sage, pp. 218–242.

Sahlin-Andersson, K. and Sevón, G. (2003). Imitation and identification as performatives. In: B. Czarniawska and G. Sevón, eds., *The northern lights: Organization theory in Scandinavia*. Malmö: Liber Ekonomi, pp. 249–265

Schmidt, S., Brinks, V. and Brinkhoff, S. (2014). Innovation and creativity labs in Berlin Organizing temporary spatial configurations for innovations. *Zeitschrift Für Wirtschaftsgeographie*, 58(4), pp. 232–247.

Shearmur, R. and Doloreux, D. (2016). How open innovation processes vary between urban and remote environments: Slow innovators, market-sourced information and frequency of interaction. *Entrepreneurship and Regional Development*, 28(5–6), pp. 337–357.

Spinuzzi, C. (2012). Working alone together: Coworking as emergent collaborative activity. *Journal of Business and Technical Communication*, 26(4), pp. 399–441.

Spinuzzi, C., Bodrožić, Z., Scaratti, G. and Ivaldi, S. (2019). "Coworking is about community": But what is "community" in coworking? *Journal of Business and Technical Communication*, 33(2), pp. 112–140.

Storper, M. and Venables, A.J. (2004). Buzz: Face-to-face contact and the urban economy. *Journal of Economic Geography*, 4(4), pp. 351–370.

Vidaillet, B. and Bousalham, Y. (2018). Coworking spaces as places where economic diversity can be articulated: Towards a theory of syntopia. *Organization*. https://doi.org/10.1177/1350508418794003.

Wenger, E. (1998). Communities of practice: Learning as a social system. *Systems Thinker*, 9(5), pp. 2–3.

# 9 Cultural intermediaries revisited

## Lessons from Cape Town, Lagos and Nairobi

*Roberta Comunian, Lauren England and Brian J. Hracs*

### Introduction

Within cultural and creative industries (CCIs), including art, music and fashion, cultural intermediaries are widely acknowledged as key players (Jakob and van Heur, 2015; Hracs, 2015). Originally defined by Bourdieu (1984) as market actors existing in-between producers and consumers, they are involved in the framing, qualification and circulation of symbolic goods, services and experiences (Maguire, 2014). These individuals share common characteristics, including high levels of cultural capital, and positions within subcultures, scenes, industries and organisations, which contribute to and validate their legitimacy and authority (Maguire, 2014, see Jansson and Gavanas, this volume). According to Bourdieu, in the 1960s, these professional taste-makers were easy to identify—producers of television programs and museum curators for example—but the range of actors, the fields they operate in, their roles and motivations have all expanded (Ashton and Couzins, 2015). Indeed, beyond individual human actors, intermediaries can be organisations, events, spaces and socio-technical actors, such as music streaming platforms (Jansson and Hracs, 2018; Hracs and Webster, 2020). The varied motivations of intermediaries have also extended to include education or preservation and to embrace more supporting and enabling roles, rather than purely working at the interface between creatives and markets/audiences (Comunian, Hracs and England, 2021).

Thus, the nature of creative labour and collaborative creative practices continue to evolve—as tension 1 in this book highlights—and situated case studies are needed to update and nuance our understanding of entrepreneurship and intermediation in the creative economy. Indeed, existing literature often describes intermediaries narrowly as co-producers, gatekeepers, brokers, agents, match makers and taste makers but the exact nature of the positions these actors hold within value chains and networks and the functions

DOI: 10.4324/9781003197065-9

they perform within the marketplace remain ambiguous (Hracs, 2015). Jansson and Hracs (2018) argue that the concept of 'cultural intermediaries' or 'cultural intermediation' has become an 'umbrella term' for a variety of actors and activities that needs further examination.

Building on the gaps identified in the literature and tensions identified by this book both on collaborative creative practices and poor representation of a range of locations, spaces and scales, this chapter presents findings from research in three African cities—Cape Town, Lagos and Nairobi—to explore intermediary practices in specific spatial contexts and industries. The aim is to probe the edges of existing conceptualisations of intermediation to see whether additional actors, activities and motivations should be included. For example, the chapter explores the work of intermediaries that do not necessarily engage with cultural content in a traditional sense—passing judgment on taste, cultural value or market value—but who enable and support the development of creative individuals and businesses. There are three key findings. First, in the contemporary creative economy cultural intermediaries come in all shapes and sizes including individuals, large formal organisations or informal community networks. Second, beyond preserving or translating cultural content, these actors provide a range of supports including space and equipment, finance, networking, training and business advice. Finally, these actors operate under different business models and exhibit diverse motivations stretching from economic profit to community building and cultural development. Ultimately, the chapter asserts that although the work of cultural intermediaries in connecting producers and consumers (Bourdieu, 1984; Maguire, 2014) remains central to many sub-sectors of the creative economy, we need to take a wider view in relation to the range of intermediaries operating and interacting in this ecosystem (Comunian, 2011).

## Cultural intermediaries in the context of change

The aforementioned expansion and extension of cultural intermediaries has been shaped by a range of changes in creative economies around the world over the past two decades. In the sections that follow, we highlight four such changes and some key implications. Crucially, there has been a *(1) redefinition of the creative sector*. From the late 1990s there has been a shift—led by policy—to move from the 'traditional' narrowly defined cultural industries (music, film, publishing) to the broader creative industries (for a critique see Galloway and Dunlop, 2007), encompassing sectors like fashion, architecture, video games and software. This expansion has also corresponded with an expansion of the intermediaries involved and their role, not just in connecting producers and consumers, but providing a complex ecology for these industries (Comunian, 2019). We have also witnessed *(2) increasing access to culture*. As levels of school attainment and income in developed and developing economies increase, we see a growth in creative and cultural consumption. This access and consumption are also facilitated by new technologies and digital platforms such as Spotify for music or Netflix for film and television (Hracs and Webster, 2020). While digital production and distribution is said to close the gap between production and consumption, the growing demand and access implies that alongside the curation of cultural content by intermediaries, we see more cultural intermediaries being involved in supporting creative careers through business advice, skills and digital platforms (Hracs, 2015; Jansson and Hracs, 2018; Comunian, Hracs and England, 2021). By extension, *(3) creative industries have come to be regarded as engines of economic growth*. Underscoring the importance of tension 3 in this book (isolated and interconnected spaces of creativity), there is not only increased policy emphasis on assessing the contributions of CCIs to economies in the Global North and established 'centres of creativity' but also in a diverse range of urban, rural, remote and emerging contexts (Comunian, Hracs and England, 2021, see also Granger, this volume; Capdevila, this volume). For example, Lai Mohammed, Nigeria's current Minister of Information and Culture, declared in 2017 at a conference on finance for the creative industries, "To those who may still be wondering, what is in the Creative Industry? My answer is that it is Nigeria's new oil. Other countries have also taken advantage of the industry to grow their economy" (Lai Mohammed, 2017). Whether this is true or more wishful thinking from policy makers, we argue the policy discourses around the CCIs have a performative impact nationally and internationally. Concomitantly, we see the *(4) defunding and marketisation of culture*. In many countries worldwide we see a reduction of the role of public funding and public policy in supporting and defining culture (Čopič et al., 2013). In this context, the role of intermediaries becomes even more central in enabling financing and allowing experimentation with new business models for culture (Monclus, 2015).

Although often overlooked, we argue that cultural intermediaries in CCIs occupy positions that allow them to connect with policy (local, national and international) as well as with key entrepreneurial and educational infrastructures within the local contexts in which they operate (Comunian, 2017a, 2017b; Comunian, Hracs and England, 2021). Indeed, Jakob and van Heur (2015, p. 357) identify various kinds of intermediaries including "arts and cultural councils, policy networks, economic development agencies, foundations and unions to artist collectives, cultural centres, creative industries incubators, festivals and tradeshows". This suggests that part of nuancing our understanding of cultural intermediaries is considering both sector specificity and the role of local context, including market dynamics and policy frameworks. It also implies acknowledging a wider ecosystem of intermediaries acting and interacting across the sector, with a broad range of objectives and motivations, providing supports and services to producers without necessarily being concerned with the cultural content of what has been produced. Although these individuals and organisations are certainly central in supporting the creative economy, unlike earlier definitions, they do not necessarily position themselves between producers and consumers in relation to translating or curating content or defining the cultural value of specific cultural artefacts. Rather, they tend to locate themselves 'alongside' producers—acting as co-producers and facilitating access to a range of resources and support services—or 'behind' producers providing finance, advice or other forms of initial training.

## Cultural intermediaries in the context of creative economies in Africa

Our research asserts the importance of looking beyond the established centres of creativity, engaging with tension 3 identified by this book. We believe there are four main reasons why creative economies in Africa provide an interesting input to the traditional understanding of cultural intermediaries in CCIs; these specifically relate to the points made above regarding contemporary changes that have expanded the role of intermediaries in new areas beyond the cultural intermediation defined in the 1960s. Although our findings are contextual to the three cities studied (Cape Town, Lagos and Nairobi), the literature and data suggest similar conditions and constraints are present more broadly across Africa (Comunian, Hracs and England, 2021; De Beukelaer, 2017). Firstly, *creative economies in Africa are reliant on the quick expansion of digital (mainly mobile) technologies*, with many young people leapfrogging analogue and traditional platforms for cultural consumption to move directly to the digital. This centrality of technology and its importance connects with key demographics of many African countries which feature exponential growth of urban youth (Sommers, 2010), corresponding with the rise of youth entrepreneurship around the exploitation of new technologies (Strong and Ossei-Owusu, 2014). Secondly, *African countries often lack formal cultural policy frameworks and institutions as well as funding for culture and creative development* (Mokuolu, Kay and Velilla-Zuloaga, 2021). Many of the formal institutions seem to be strongly anchored to traditional sectors of the economy (extraction, agriculture, manufacturing) and less concerned with emerging sectors. Schneider and Gad (2014) talk about a weak role for cultural policy in the context of Africa; for example, Obia et al. (2021) discuss the inability of Nigeria to redraft its cultural policy in 2008 even though the original cultural policy framework of 1988 can be considered inadequate for modern times. Thirdly, despite the lack of funding (Mokuolu, Kay and Velilla-Zuloaga, 2021) and formal recognition of CCIs as a sector in many countries, *the economic impact of the sector is often staggering and has been considered key to economic development by many international organisations* (UNCTAD, 2010; UNESCO and UNDP, 2013). In 2012, UNESCO rated Nigeria's film industry as the world's second largest (Comunian and Kimera, 2021), highlighting the role of the sector across Africa and globally. Finally, and most importantly, many authors studying creative economies in Africa highlight *the role of informality and networks* in the industry structure (Bello, 2021). With respect to the Tanzanian film industry, Mhando and Kipeja (2010) highlight how local film-makers and producers are often isolated and rely on micro and informal enterprises, without being able to access networks that could enable them to enter global markets. These issues suggest that intermediaries are able to broker relations and support CCIs and that they play a vital role in the development of the sector.

## Methodology and data

The chapter presents findings from a research project entitled 'Understanding and Supporting Creative Economies in Africa: Education, Networks & Policy'[1] (2016–2019) funded by the AHRC (Arts & Humanities Research Council, UK). It aimed to map creative economies across Africa but given time and resource limitations we selected three countries (Kenya, Nigeria and South Africa), representing the context of West Africa, Southern Africa and East Africa more broadly. The fieldwork is focused on large cities in these countries, one capital (Nairobi) and two crucial centres of creative production (Cape Town and Lagos). We acknowledge that these cities are not representative (in relation to scale and importance) of most African cities, however, they provided an interesting ground for reflection in relation to the importance of creative economies in their international profile and urban dynamics. The intensive fieldwork, with a week spent in each city, allowed us to collect a range of data. Across the three cities this included a total of 29 focus groups, 35 interviews (some undertaken via Skype), participant observation, 13 research visits to local arts/creative organisations and an online survey (85 respondents) with cultural intermediaries working in those cities. This chapter focusses on the analysis of the data from the online survey, but the discussion benefits from reflection on the broader fieldwork undertaken. The survey was disseminated via social media platforms from the project accounts (Twitter, Facebook and Instagram), mailing lists (including local partners' contact lists), policy organisations (such as the British Council) and other cultural agencies (such as the South African Cultural Observatory) for a period of three weeks for each city. The survey aimed to reach as many individuals as possible in the local context and also offered individuals the possibility to get further involved by attending a focus group on this topic. The survey provided a broad definition of cultural intermediaries and asked participants that self-identified with the definition to complete the survey. The survey included both closed (nature of the organisation, location, field of work etc.) and open-ended questions, allowing participants to reflect on the challenges they faced and their relationships with other intermediaries. The Lagos survey received 32 responses; the Cape Town survey received 25; and the Nairobi survey received 28. The respondents belonged to a range of organisations (see next paragraph) and tended to be professionals in a range of leadership roles. The survey and data collection had limitations with respect to its self-selective nature. We were not able to establish the representative value of the sample as there is no available data, to our knowledge, on the overall population of intermediaries. However, the survey allowed individuals to add activities and reflect on what their role in the sector was, which was our main interest.

## Who are the intermediaries in Cape Town, Lagos and Nairobi?

Our data shows that as a term 'cultural intermediaries' was accepted by respondents to include a range of actors. Despite the differences across cities and countries that can be seen in Table 9.1, what emerges clearly from the survey is that cultural intermediaries can be individuals or organisations in a range of sectors of the economy with a range of business models, including individual freelancers, large public sector organisations and informal community networks. These business models are accompanied by a range of motivations stretching from economic profit to community building and support, cultural development and equality or inclusion. From the data we can imply that in countries with different cultural policy frameworks—such as Nigeria and South Africa—there might be an emphasis on certain business models or opportunities, for example the higher proportion of NFP/Charities in Cape Town compared to more private companies in Lagos and Nairobi. However, overall the profile of cultural intermediaries across Africa highlights the importance of looking across a range of sectors. The respondents also covered a range of jobs with a variety of job titles. Overall there was a higher proportion of intermediaries identifying as Founders/Directors/CEOs (51% total Lagos; 41% CT; 43% Nairobi).[2]

*Table 9.1* Cultural intermediaries and organisational typology

| *Organisation/ Company Type* | *Lagos* | | *Cape Town* | | *Nairobi* | |
|---|---|---|---|---|---|---|
| | *Number of Intermediaries* | *%* | *Number of Intermediaries* | *%* | *Number of Intermediaries* | *%* |
| Individual/ Freelancer | 1 | 3% | 3 | 12% | 3 | 11% |
| Informal/ Community network | 2 | 6% | 2 | 8% | 2 | 7% |
| Not-for-profit/ Charitable organisation | 4 | 13% | 14 | 56% | 5 | 18% |
| Private company | 17 | 53% | 4 | 16% | 13 | 46% |
| Public sector organisation | 8 | 25% | 2 | 8% | 5 | 18% |
| Total | 32 | | 25 | | 28 | |

## What do the intermediaries in Cape Town, Lagos and Nairobi do?

We were very interested in exploring the range of activities that cultural intermediaries performed in the CCIs in these cities. Table 9.2 articulates the supports they offer under a range of headings.

The open questions of the survey allowed us to get a better understanding of this provision and its context and importance. Based on our data and wider research, we present this analysis and the provision in five main support categories: space and equipment (n=45, 53%); finance (n=34, 40%); networking and partnership (n=60, 71%); training and skills (n=64, 75%); business guidance and advice (n=61, 72%). Most acted in at least two of these provisions, confirming that in general actors have multiple roles and positionalities (Jansson and Hracs, 2018), only 8% (n=7) offered only one service.

### *Space and equipment*

Providing examples of the ‘spaces of creativity’ outlined in tension 3, this broad category includes a range of provisions, such as office space, making space, coworking space, and business incubation as well as rehearsal/studio or performing spaces, including access to specialised equipment and IT infrastructure. For the initial development of CCIs space and equipment are very important. Small start-ups or even freelancers cannot cover the costs of office space and making facilities, therefore the

*Table 9.2* Services and support provide by cultural intermediaries

| *Services/support offered* | *Lagos* | | *Cape Town* | | *Nairobi* | | *Total* | |
|---|---|---|---|---|---|---|---|---|
| | *Number of Intermediaries* | *%* | *Number of Intermediaries* | *%* | *Number of Intermediaries* | *%* | *Number of Intermediaries* | *%* |
| Finance | 16 | 50% | 14 | 56% | 4 | 14% | 34 | 40% |
| Finance and business finance support | 8 | 25% | 6 | 24% | 1 | 4% | 15 | 18% |
| Funding/Commissioning of arts/creative projects | 15 | 47% | 10 | 40% | 4 | 14% | 29 | 34% |
| Business advice and guidance | 26 | 81% | 17 | 68% | 18 | 64% | 61 | 72% |
| Business advice | 14 | 44% | 9 | 36% | 9 | 32% | 32 | 38% |
| Guidance and policy advice | 11 | 34% | 10 | 40% | 8 | 29% | 29 | 34% |
| Research and consultancy services | 21 | 66% | 10 | 40% | 9 | 32% | 40 | 47% |

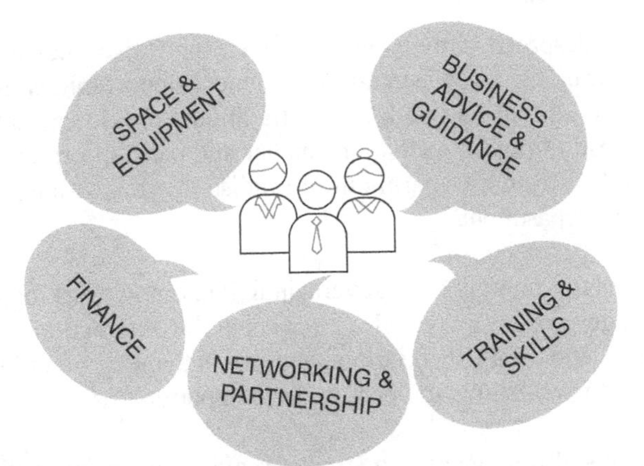


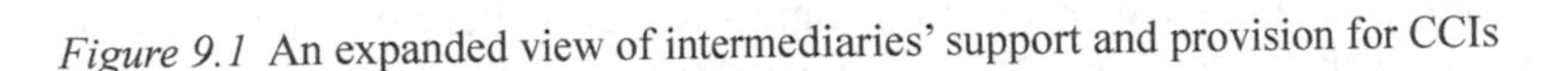

*Figure 9.1* An expanded view of intermediaries' support and provision for CCIs

opportunity for intermediaries—often creatives themselves– to maximise the use of shared spaces and equipment benefits the sector enormously. However, space and equipment are often associated with further provision such as networking or brokering new opportunities for CCIs. As one respondent from Lagos explained: "The objective of the exchange is to mobilise talented individuals and creative entrepreneurs to form the biggest Creative Hub in the North-Central Region of Nigeria, through cross-disciplinary joint ventures or group enterprises". Space and equipment are therefore used as opportunities to create critical mass and maximise limited resources.

### *Networking and partnerships*

Highlighting the collaborative nature of creative practices (tension 1 in this book), here we include organising events, matching and brokering, partnerships and collaborations, festivals, tradeshows as well as specialised networking opportunities or brokering. The networking and brokering flows in all directions; it can connect to other producers, to consumers or to other intermediaries. As one respondent from Nairobi puts it: "We are part of the Creative Economy Working Group, a consortium of 16 organisations who come together to promote creativity, innovation and the diversity of the cultures of the people of Kenya as drivers of economic growth and development". In this space there is a broader agenda to create opportunities that stretch from commercial to community-oriented goals.

### *Training and skills*

This is a widespread activity and includes specialised business training or creative training, professional development opportunities and mentoring for CCIs. This occurs both within traditional providers such as higher education institutions as well as in smaller and more informal settings. As one respondent from a Department of English in a Lagos-based institution argued, their department

> nurtured the emergence of several individuals now working in the creative industries . . . created . . . avenues for interaction between our students and established writers and works to mentor and link aspiring writers to publishing and performance opportunities.

However, this also happens at a smaller scale for other intermediaries; for example, as the artist-owner of a Cape Town art gallery explains: "we run art workshops for rural Mpumalanga artists to improve their skills and enhance their techniques". Therefore, irrespective of the size and formality of how training is delivered, many intermediaries provide learning opportunities for other creatives within their practices and work.

### *Business guidance and advice*

This can include guidance on start-up, finance and growth as well as export and internationalisation, IP and market access. As one respondent from Cape Town explained: "we support the development of small creative businesses operating in the craft and design sector—working closely with the entrepreneurs on developing business knowledge, developing their products, and facilitating producers with access to market". In many cases there was an emphasis for this business guidance to be supported or provided by public sector bodies for the potential it has to benefit the economy more widely. However, due to the lack of policy frameworks and funding, it was often left to not-for-profit or commercial organisations to deliver, limiting the availability or accessibility of these opportunities.

### *Funding and finance*

This includes distribution of public funding or from trusts; repayable loans and non-repayable grants; finance/investment brokering as well as crowdfunding. There are few organisations providing finance solely to creatives and in most cases finance came within a broader business development and support framework (especially when loans and return on investment are needed). As one respondent from Lagos told us:

> finance is critical to our effectiveness. Hence our training and networking should emphasise public funding, loan acquisitions, crowdfunding etc. Part of our role as a creative intermediary is to encourage creative ways of achieving the above, by leveraging social media tools especially crowdfunding tools.

Therefore, the cultural intermediaries might provide funding but also broker or provide skills to access finance for creatives.

All the activities, services and expertise discussed by our participants highlight how the activities of cultural intermediaries in CCIs stretch across a wide spectrum of work. While in many of these roles an element of curation (Jansson and Hracs, 2018) or traditional cultural intermediation might still be present, overall we found that the main agenda of many was not to preserve, translate or curate the cultural content proposed, but to simply allow creatives (from music and film to fashion and craft) to reach their potential and make a sustainable living from their creative products or practices.

It is important to note that the traditional view of intermediaries addresses the relationship between producers and consumers or the market and in many ways assumes a post-production interaction when cultural and creative products are evaluated, promoted, sold or cared for. There has to date been little consideration of the support that is needed for pre-production stages including conceptualisation, design, development and creation (skills, mentorship, space, finance etc.). This research suggests that many of the services, supports or interactions delivered or facilitated by intermediaries occur at different stages in the development journey. For example, education can be the starting point or something creatives return to, finance might be needed early on or after an initial period of business development and space requirements may change over time. A key finding is therefore not just to identify an expanded spectrum of activity, performed by a range of actors with a range of motivations, but also to understand that these activities are performed at different stages. By extension, while these activities may be shared by intermediaries operating in CCIs around the world, it is clear that industrial and contextual specificity, including local politics, infrastructure provision and market dynamics, play a key role in shaping these practices and generating a range of challenges that must be negotiated. Underscoring the importance of tension 3, the experiences of cultural intermediaries in our study varied by city, industry and scale and are arguably very different from those of their counterparts in cities in the Global North.

## Conclusion

Despite the tendency within existing literature to focus on the experiences of individual creatives who work in isolation, this chapter engaged with tension 1 by exploring the collaborative dynamics of creative practices and nuancing our understanding of the roles that intermediaries play in enabling, driving and supporting CCIs more broadly. By drawing on case studies of Cape Town, Lagos and Nairobi the chapter also provided insights into three emerging but poorly understood creative economies contexts (tension 3).

The findings from the survey highlight that cultural intermediaries include a very broad set of individuals, actors, organisations, spaces and digital platforms that aim to facilitate the growth and development of creative individuals and CCIs. We discovered that they play a vital role in supporting the sector by providing access to information, skills, resources and networks that enable them to fulfil their mission, creative or business goals. Rather than positioning themselves exclusively between producers and consumers or audiences, they assume a range of positions along the supply chain, from providing training earlier on in creatives careers, to sitting alongside producers during projects as co-producers.

Our findings also revealed that cultural intermediaries come in all shapes and sizes but also have different motivations. Some are passionate individuals with experience in the creative and cultural sector who provide training and mentorship, others are large publicly-funded organisations with the remit to distribute funding and provide feedback to artists. We also discovered that these intermediary actors operate under a range of business and organisational models including for profit, public-sector organisations, charities or not-for profit organisations, cooperatives and social enterprises or through informal networks. These actors can operate in a range of sectors, providing a variety of services or specialising in just one of them. They can provide access to information and networks (soft infrastructure) as well as physical or structural resources (hard infrastructure). The five areas of CCI support we have identified are: providing space and equipment; supporting access to finance; facilitating networks and partnerships; offering opportunities for training and development; and providing specialised business guidance.

Overall this chapter aimed to underscore the importance of the creative sector and the need for a better understanding of the role that cultural intermediaries can play in the development of creative economies in Africa and around the world. Yet, ongoing research is needed to provide a more detailed picture of how these actors operate in relation to policy and development agendas—especially with respect to access and equality. Given the importance of specificity we were reluctant to simply import the concept of cultural intermediaries from the Global North into the Global South. Moreover, while the concept resonated with actors in Africa and remained useful in exploring the emerging creative economies of Cape Town, Lagos and Nairobi, it is clear that it can and should be challenged to be more inclusive of different policy agendas and developmental stages. Ultimately, a new term which embraces and more accurately conceptualises the range of actors, roles, motivations and objectives might be needed.

## Notes

1 We acknowledge the support received by the AHRC (Arts & Humanities Research Council) UK grant number AH/P005950/1 (2016–2019). More information on the project can be found at: www.creative-economy-africa.org.uk.

2 Intermediaries across the three cities identifying as Founder/CEO or Founder/Director were coded as 'Founder' (12%); 'Academic' code includes 'Lecturer', 'Head of Dept' and 'Researcher' (13%); 'Director' includes 'Artistic Director', 'Programme Director' and 'Executive Director' (26%).

## References

Ashton, D. and Couzins, M. (2015). Content curators as cultural intermediaries: 'My reputation as a curator is based on what I curate, right?' *M/C—A Journal of Media and Culture*, 18(4).

Bello, M. (2021). Collaborations for creative arts higher education (HE) delivery in Ghana. In: R. Comunian, B.J. Hracs, and L. England, eds., *Higher education and policy for creative economies in Africa: Developing creative economies*. London: Routledge, pp. 28–43.

Bourdieu, P. (1984). A Social Critique of the Judgement of Taste. In: *Traducido del francés por R. Nice*. Londres: Routledge.

Comunian R. (2011). Rethinking the creative city: The role of complexity, networks and interactions in the urban creative economy. *Urban Studies*, 48, pp. 1157–1179.

Comunian, R. (2017a). Creative collaborations: The role of networks, power and policy. In: M. Shiach and T. Virani, eds., *Cultural policy, innovation and the creative economy: Creative collaborations in arts and humanities research*. London: Palgrave Macmillan, pp. 231–244.

Comunian, R. (2017b). Temporary clusters and communities of practice in the creative economy: Festivals as temporary knowledge networks. *Space and Culture*, 20(3), pp. 329–343.

Comunian, R. (2019). Complexity thinking as a coordinating theoretical framework for creative industries research. In: S. Cunningham and T. Flew, eds., *A research agenda for creative industries*. Cheltenham: E Elgar.

Comunian, R., Hracs, B.J. and England, L. (2021). *Higher education and policy for creative economies in Africa: Developing creative economies*. London: Routledge, pp. 60–77.

Comunian, R. and Kimera, G. (2021). Uganda film and television. In: R. Comunian, B.J. Hracs, and L. England, eds., *Higher education and policy for creative economies in Africa: Developing creative economies*. London: Routledge, pp. 60–77.

Čopič, V., Inkei, P., Kangas, A. and Srakar, A. (2013). Trends in public funding for culture in the EU. Brussels, European Union: EENC report.

De Beukelaer, C. (2017). Toward an 'African' take on the cultural and creative industries? *Media, Culture & Society*, 39(4), pp. 582–591.

Galloway, S. and Dunlop, S. (2007). A critique of definitions of the cultural and creative industries in public policy. *International Journal of Cultural Policy*, 13(1), pp. 17–31.

Hracs, B.J. (2015). Cultural intermediaries in the digital age: The case of independent musicians and managers in Toronto. *Regional Studies*, 49(3), pp. 461–475.

Hracs, B.J. and Webster, J. (2020). From selling songs to engineering experiences: Exploring the competitive strategies of music streaming platforms. *Journal of Cultural Economy*, 14(2), pp. 240–257.

Jakob, D. and van Heur, B. (2015). Editorial: Taking matters into third hands: Intermediaries and the organisation of the creative economy. *Regional Studies*, 49(3), pp. 357–361.

Jansson, J. and Hracs, B.J. (2018). Conceptualizing curation in the age of abundance: The case of recorded music. *Environment and Planning A: Economy and Space*, 50(8), pp. 1602–1625.

Lai Mohammed, A. (2017). *Press conference by Alhaji Lai Mohammed, on creative industry financing conference*. Federal Ministry of Information and Culture, Nigeria. Available at: https://fmic.gov.ng/press-conference-alhaji-lai-mohammed-creative-industry-financing-conference/.

Maguire, J.S. (2014). Bourdieu on cultural intermediaries. In: *The cultural intermediaries reader*. London: Sage, pp. 15–24.

Mhando, M.R. and Kipeja, L. (2010). Creative/cultural industries financing in Africa: A Tanzanian film value chain study. *Journal of African Cinemas*, 2(1), pp. 3–25.

Mokuolu, Y., Kay, V. and Velilla-Zuloaga, C.M. (2021). Finance for creative and cultural industries (CCIs) in Africa. In: R. Comunian, B.J. Hracs, and L. England, eds., *Higher education and policy for creative economies in Africa: Developing creative economies*. London: Routledge, pp. 113–130.

Monclus, R.P. (2015). Public banking for the cultural sector: Financial instruments and the new financial intermediaries. *International Review of Social Research*, 5(2), pp. 88–101.

Obia, V., England, L., Comunian, R. and Oni, D. (2021). Creative higher education (HE) in Nigeria and the case of university of Lagos. In: R. Comunian, B.J. Hracs, and L. England, eds., *Higher education and policy for creative economies in Africa: Developing creative economies*. London: Routledge, pp. 11–27.

Schneider, W. and Gad, D. (2014). *Good governance for cultural policy: An African-German research about arts and development*. Frankfurt am Main: Peter Lang.

Sommers, M. (2010). Urban youth in Africa. *Environment and Urbanization*, 22(2), pp. 317–332.

Strong, K. and Ossei-Owusu, S. (2014). Naija boy Remix: Afroexploitation and the new media creative economies of cosmopolitan African youth. *Journal of African Cultural Studies*, 26(2), pp. 189–205.

UNCTAD (United Nations Conference on Trade and Development). (2010). *Creative economy report 2010: Creative economy—a feasible development option*. Geneva. Available at: https://unctad.org/system/files/official-document/ditctab20103_en.pdf.

UNESCO (United Nations Educational, Scientific and Cultural Organisation) and UNDP (United Nations Development Programme). (2013). *Creative economy report 2013—special edition: Widening local development pathways*. New York. Available at: www.unesco.org/culture/pdf/creative-economy-report-2013.pdf.

# 10 Curated by pioneers, spaces and resistance

## The development of electronic dance music in Stockholm

*Johan Jansson and Anna Gavanas*

### Introduction

Stockholm has been recognised as an internationally influential music scene and an important node in the international music industry, producing influential artists in various genres, song writers, music instruments and music industry related services (Power and Jansson, 2004; Hracs and Webster, 2020; Johansson, 2020) This creative music environment has been studied from a variety of perspectives emphasising institutional, network and innovation-based explanatory models (Power and Jansson, 2004; Power and Hallencreutz, 2007; Johansson, 2020). While this literature generally tends to focus on popular music, this chapter describes a partially hidden piece of history—the story of how electronic dance music was introduced in Stockholm. It focusses on the development of electronic dance music in Stockholm by comparing two contrasting eras: the left-oriented and anti-commercial cultural politics of the 1970s and the neoliberal politics of the 1990s. In general, music performed at discos and dance venues (sometimes referred to as 'Gebrauchsmusik' or 'utility music') has largely been excluded from the privileged position enjoyed by other popular music in the music industry, media and cultural policy (Thornton, 1995; Wilderom and van Venrooij, 2019). Nevertheless, these genres have been vital ingredients of the national music scene in and outside Stockholm as well as being a significant influence on the relative success that Swedish music production and artists have experienced over the last 50 years. From the disco ABBA produced in the 1970s, through SweMix house in the 1980s, to the Swedish House Mafia from the 2000s and on, Stockholm has been a centre for electronic dance music as well as an important context where pioneering, and (re)fashioning of cultural expressions and musical genres, has taken place (Gavanas and Öström, 2016; Gavanas, 2020).

Empirically, this chapter utilises longitudinal and historical perspectives of the first two waves of electronic dance music (disco and rave) in

DOI: 10.4324/9781003197065-10

Stockholm and Sweden (also presented in Gavanas, 2020). From the disco era, the chapter draws on material from the main Stockholm based newspapers 1974–1985 as well as magazines for professional DJs 1978–1981. From the rave era, the chapter considers material from the Rave Archive (privately donated by the former scholar Elisabeth Tegner). The archive contains 145 articles, hundreds of flyers and 103 interviews with participants (DJs, club organisers and clubbers/ravers) of the electronic dance music scene in Stockholm—as well as field notes from participant observation 1988–1995. Furthermore, archival material is supplemented with 40 retrospective interviews with DJs, promoters and key participants for both the disco and rave eras (Gavanas and Öström, 2016).

To analyse this material and understand the development of electronic dance music in Stockholm this chapter uses the concept of curation. While the concept of curation traditionally emphasises the role of intermediary actors in sorting, filtering and creating value (Balzer, 2014), more recent literature is broadening this understanding. Indeed, growing interest has been paid to the motivations and spatial dynamics associated with curatorial processes (Joosse and Hracs, 2015; Hracs and Jansson, 2020; Hracs and Webster, 2020; Merkel and Suwala, this volume). The chapter adds further nuance to our knowledge of curation by examining three interrelated and theoretically informed themes. The first theme focuses on translocal flows and highlights the role pioneers play in bringing electronic dance music to Stockholm and developing and simultaneously re-interpreted cultural expressions in different local/global locations. The second theme focusses on how the local milieu provides spaces, or an infrastructure, for electronic dance music to develop, such as nightclub scenes, public youth recreation centres and rave parties. These constitute important (urban) spaces of curation. In the third theme, the role of resistance is analysed. We look at how individuals, collectives and spaces formed and formulated resistance against policy (e.g. regulations) and the mainstream media (of the early 1970s to the mid-1990s).

Thus, the aim of this chapter is twofold. Empirically, the chapter emphasises periods and genres that are rarely recognised in the music industry literature. A second aim is to contribute to the literature on curation by showing how time and space-specific processes in Stockholm are curating our understanding and experience of specific (sub)cultural expressions and thus the development of electronic dance music in Stockholm. In relation to the tensions articulated within this book, the chapter discusses spatial processes underlying the development of (sub)cultures by highlighting translocal flows of products and knowledge, i.e. the migration of ideas and processes from cultural epicentres to peripheral places (such as Stockholm, see also Capdevila, this volume). Interactions and tensions existing within urban spaces are also discussed; between socio-economically distinct spaces of the inner city and the suburbs. Thus, this chapter connects to the tension between isolated and interconnected spaces of creativity, as well as the tension between individual and collaborative creative practices.

## Pioneers bringing electronic dance music to Stockholm

Translocality is a concept that describes how different phenomena are spread across multiple sites in networks and how different sites are linked thanks to these networks (Greiner and Sakdapolrak, 2013). This concept rejects a strict division between local, regional and global scales, and instead looks at how things that happen locally also affect and are influenced by similar processes and phenomena in other places. Thus, translocality is a way to identify increasing social complexity brought about by processes taking place in global, inter-regional and digital networks. Translocality could be defined as "being identified with more than one location" (Oakes and Schein, 2006, p. 1) or a polycentric phenomenon (Maly and Varis, 2015). While some attributes are found globally, other attributes are more context-specific and as such are 'translated' in local contexts. Consequently, there are multiple centres for 'normative orientation' (Maly and Varis, 2015, p. 644). In relation to electronic dance music, Gibson (1999) recognises the global connectivity of rave culture and Thornton (1995, p. 3) argues that, while club cultures on the one hand are a global phenomenon, it is simultaneously embedded in local practices; "Dance records and club clothes may be easily imported and exported, but dance crowds tend to be municipal, regional and national".

Our empirical material supports the idea that specific individuals act as pioneers; early adopters of new music and ideas, such as DJs and promotors. These individuals constitute nodes in international (and national) flows that introduce music and organisational ideas from other places (mainly UK and USA) and in translating (Latour, 1987) different aspects of disco and rave to national and local circumstances and regulatory frameworks.

From the early 1970s until the mid-1990s, the first waves of disco and house/techno were introduced to Stockholm through translocal flows of traveling pioneers, music (radio, magazines and vinyl records) and styles (dances, aesthetics and fashions). Pioneering Swedish DJs and club promoters travelled to the US and UK and were inspired by the new sounds, styles and dance cultures. Commonly, these pioneers came from resourceful contexts with the means and time to travel or connect through networks. For instance, from the late 1960s and onwards, DJ and club promoter Claes Hedberg worked on various ships between the UK and the Swedish west coast. In addition, his uncle was an airline pilot who brought records from the US which had not yet been released in Sweden. Claes also gained interest and knowledge when visiting *Studio 54* in New York and was DJing at nightclubs in Stockholm and elsewhere in Sweden. From his travels, contacts with record labels and import shops, he had access to records and in the late 1970s and early 1980s he also ran a Swedish national organisation for DJs named SYD (Sweden´s Professional DJ association), organised national meetings (Discoforum 1979–1984) for Swedish DJs, and promoted disco music nationally.

Likewise, other disco pioneers in Stockholm, like Alexandra Charles (*Alexandra's* Nightclub) and Sydney Onayemi (*Big Brother*), conceived of their night club concepts through international travel and networks. Disco pioneers followed a long tradition of international flows of dance music scenes to Stockholm, dating back to the jazz age of the 1940s. During the pre-disco music era of the 1960s, pioneers like Eva Carlsson and Anette Hedlund went from Stockholm to France, visited discotheques like *La Gorille* in Nice and *Whiskey a go go* in Paris and brought back French and Italian records. Alternative sources for inspiration were international radio channels (e.g. Radio Luxembourg), printed magazines and imported records.

Also, in the domestic Swedish context, knowledge and organisational skills about dance music cultures were re-interpreted and diffused through pioneering individuals where it spread between suburbs and city centres—as well as between different cities in Sweden. One such pioneer, Jan Palm was a DJ and promoter in Stockholm in the 1970s and 1980s. He lived in the Stockholm suburb Tensta, where he first discovered disco though the multilingual commercial broadcaster Radio Luxembourg.

In the 1980s the popularity of disco faded, however, when mutating into house music similar developments can be discerned. The introduction of house music in Sweden at the end of the 1980s was led by a limited but enthusiastic group of individuals that had access to products and knowledge from various sources. For example, through the record shop *Vinylmania* in Stockholm, René Hedemyr had discovered house music in 1985 (subsequent to being one of Stockholm's disco pioneers since the early 1970s). The owner John Wallin gave René a bunch of records as well as an issue of *GQ* magazine featuring an article on Chicago House. As a result, René became a house music curator and organised, with his *Swemix* crew, one of Stockholm's first house clubs called *Bat Club* at the *Ritz* nightclub on Medborgarplatsen in Stockholm.

Swedish DJs and promoters who traveled to the UK in the late 1980s tapped into the musical and cultural flows when acid house parties transformed into rave parties in warehouses and outdoor spaces. The *Bat Club* became a bridge between house and rave and the DJ, club promoter and producer Mårten Attling, an early rave pioneer in Stockholm, was influenced by the acid house movement, also known as the "second summer of love" in London in 1988. In 1990 Mårten arranged the rave party *True London Style Rave* in Kungsängen outside of Stockholm. For Mårten, arranging his own rave party was also the first rave party he ever attended and in general, the rave culture initially was characterised by a passionate DIY attitude. The organisers did not have huge record collections and rather they made music themselves that they mixed with purchases from their latest trip to the UK. Also, rave collectives collaborated and formed translocal communities within international and domestic contexts. For instance, in 1991 rave promoter Mikael Eklöf and the *Space Ravers* collective arranged *Techno Woodstock*, a three-day festival in a large building in suburb Stockholm involving techno acts from Gothenburg and Malmö in Sweden as well as Denmark and Germany.

## Spaces for electronic dance music in Stockholm

Few activities are as agglomerated as the cultural and creative industries (Power and Scott, 2004). There are multiple reasons for this, but generally they relate to the 'hypersocialised production system' (Currid and Connolly, 2008) that affects (the often) project-based production (Grabher, 2002), and informal labour market processes (Caves, 2000). In addition, agglomerations also facilitate proximity to financing (Hesmondhalgh, 2002), market and consumer responses valuable in value creation processes (Crewe and Beaverstock, 1998), as well as knowledge creation and diffusion (Lorenzen, 2018). In these hyper-socialised and project-based activities embedded local milieus function as scenes or arenas for processes of evaluation and value creation, deciding which music and clubs are fashionable or not (Thornton, 1995).

Although rarely emphasised in the literature, the cultural and creative industries are dependent on the supply of appropriate and low-cost spaces facilitating socially embedded processes of creativity and valuation. From the sparse literature in the field we know that some urban spaces and places are better equipped to host cultural production and creativity than others (Molotch and Treskon, 2009). The demand for cheap and functional premises is specified partly through the inability to pay high rents, and partly because some (e.g. concert venues and discotheques) need spaces that do not disturb neighbours and other ongoing activities. Dancing, dance culture and even more specifically, club culture are territorial phenomena in such a way that they are "persistently associated with specific space which is both continually transforming its sounds and styles and regularly bearing witness to the apogees and excesses of youth culture" (Thornton, 1995, p. 3).

In analysing the empirical material and the role of built infrastructure one needs to differentiate between the *discotheques/discos* (i.e. spaces where the audience dances to recorded music) that reached Sweden during the 1960s—and the *disco music* that reached Sweden in the 1970s. Discotheques and DJs were already a nationwide phenomenon in Sweden at the end of the 1960s and thus provided the infrastructure for the introduction of electronic dance music in the 1970s. André Lahovary, a Stockholm DJ 1968–1980, recalls that there were around 20 privately owned youth discotheques in Stockholm Old Town at the end of the 1960s. These early discotheques featured live and recorded music influenced by UK pop culture for people 16–20 years old.

When disco music entered Stockholm in the 1970s, via soul and funk music, it was played at discotheques and nightclubs, but also at municipal youth recreation centres. For example, just like Jan Palm and his crew, many legendary DJs in Sweden, from the 1960s and on, started out their careers at youth recreation centres. This was also the case of René Hedemyr from Jan's Tensta crew who later became one of the first curators of the house music scene in Sweden.

Rave culture differentiated itself from nightclub culture in terms of space and infrastructure. Set in unconventional outdoor and industrial settings, rave parties were less restricted by physical and geographical limitations. Raves could be located anywhere inside and outside of the city centre and participants made a point to claim previously 'un-danced' spaces. In line with the general DIY approach, rave-organisers build sound, light and décor from scratch. In other words, whereas disco was part of the public and formalised discotheque/nightclub industry—raves were partly launched as informal alternatives to this industry (in terms of space and organisation) by loosely organised networks of supporters (as opposed to employees).

The ramifications of electronic dance music cultures were also shaped by developments in technology and communication (Hodkinson, 2004). In a retrospective interview, DJ and participant Sussi Zällh observed tendencies towards exclusivity in the late 1980s and early 1990s house and rave clubs in Stockholm, which were shaped by the physical limitations of the information channels. Prior to internet and social media, discotheques and nightclubs were advertised publicly in newspapers, whereas house, techno clubs and rave events were advertised through word of mouth in networks of friends, promoters and DJs as well as printed flyers in specialised record stores or trendy cafés/shops. Zällh furthermore observed that, by the mid-1990s, the Stockholm house/techno/rave scene had transformed, from an underground phenomenon for an exclusive crowd who was 'in the know', into a hugely popular mainstream youth phenomenon.

*Figure 10.1* Front and back of flyer for UK-inspired discotheque promotion based in Stockholm, 1970

## Resistance and the pleasure of opposition

The introduction of new ideas and cultural expressions does not come without resistance. In this case, arguably, this resistance takes two forms, which are both important to the contextualisation and development of electronic dance music. The first is 'institutionalised' resistance by forces in industry and policy. For instance, the so called 'dance permit' law which dates back to 1956, and is still enacted in Sweden, was a result of the moral panic that followed the introduction of jazz music and related culture. According to this legislation, a permit is required in order to organise public dance events in public as well as private settings. The second is the 'grassroots' resistance to this institutionalised power and the established cultural norms of mainstream society through club cultures.

Much of the literature on popular music studies are focusses on 'youth subcultures' in terms of simultaneous compliance and resistance to domination (Hall and Jefferson, 2006). For instance, hip hop, 'protest rock' and feminist/lesbian Riot Grrrls have been discussed in these terms. In contrast, electronic dance music cultures are often perceived as non-political, existing solely for the momentary and hedonistic pleasures of its participants (Buckland, 2002). In the context of New York's queer club history, Buckland opposes the constructed dichotomy between club scenes and politics and contests the idea that dance clubs produce nothing useful politically other than constituting 'dangerous' and hedonistic destruction. Instead, "struggles for the right to dance" are seen as extremely important battles in contexts where local governments seek to regulate the activities and access of certain groups in specific areas (Buckland, 2002, p. 92). In this sense, DJs are outlaws at the centre of sometimes law-breaking techno movements (Brewster and Broughton, 1999). Balliger (1995, p. 24) suggests that "temporary autonomous zones" and "sonic squatting" are "non-spatial" ways to "reclaim space". Buckland (2002, p. 144) describes club culture and queer 'world-making' in New York as "part oppositional, part pleasure, and in addition, part the pleasure of opposition". Thus, this "pleasure as politics" approach may function as an alternative and complement to the youth subculture paradigm of resistance (Balliger, 1995, p. 21).

In the 1970s and 1980s the four major Stockholm newspapers were governed by left-oriented critiques towards disco culture, e.g. "Sweden is invaded. USA is taking over with mass culture as its weapon" (Peterson, 1980). When disco music and culture was first introduced in Sweden, it was initially described through an US 'imperialist' discourse and was generally interpreted by media, policymakers and the cultural establishment in terms of a superficial, bourgeois and commercial phenomenon, void of any meaning or value. In other words, while disco was associated with marginalised queers in its original US context, its queer political implications were lost on Swedish music journalists.

Instead disco was re-interpreted according to the terms of the Swedish cultural debate at the time (Gavanas, 2020). These negative attitudes correspond to the introduction of other musical genres in Sweden, as well as in other countries (Cresswell, 2006). For example, by the 1940s, jazz music and foxtrot dance culture had been introduced to Sweden, these cultural expressions faced massive critique by Swedish authorities, and public debate adversely discussed these new cultural phenomena in terms of a foreign threats to domestic culture, youth and moral order. As with the case of jazz culture that was later re-interpreted and fully assimilated and part of a wider high cultural canon in Sweden, disco culture also conquered dancefloors all over Sweden and has been a recurring pop cultural phenomenon ever since.

Since the 1960s, Swedish cultural policy has been a major factor providing, as well as restricting, space for cultural expressions. Music schools, youth recreation centres and study associations have operated as state and municipal funded sites for fostering youth. In the 1970s, an outspoken aim for Swedish cultural policy was to counteract, and provide alternatives to, commercialism as well as promoting domestic music and musicians. Consequently, Swedish cultural policy, as well as youth policy, framed disco in terms of a hollow and profit driven machine aesthetic, with harmful effects on Swedish society.

In contrast to disco culture, 1990s rave culture was never assimilated into either the night club industry or state-funded activities for youth. Rave culture originated in underground and DIY activism—although it had no clear-cut political agenda except for its vague PLUR (Peace, Love, Unity, Respect) philosophy.

One important difference between the ways in which disco was received and interpreted in Sweden, as compared to rave, stems from the contrasting political contexts of the left-oriented 1970s compared to the neoliberal 1990s. In the market-oriented 1990s, Stockholm policymakers promoted entrepreneurship in its cultural policies. Thus, rave culture became an anomaly by defining itself as alternative to the nightclub industry as well as the physical limitations and norms of night club culture.

Rave music and culture were immediately associated with drugs and drug use when first noticed by Swedish media and the political consensus on the zero tolerance on drugs was another important factor to the resistance to rave culture by Swedish policymakers in the 1990s. After rave culture had been around for more than half a decade, the 'Rave Commission' was founded in 1996 by the Stockholm police. The large-scale rave club *Docklands* opened at the end of 1995 in an industrial building in Nacka and attracted thousands of ravers. Organisers from the *Docklands* collective openly challenged the authorities, which sparked a full-scale conflict between the authorities and rave culture. The media fuelled this fire with headline stories on raves involving cynical organisers selling drugs to school children, as well as quotes from parents worrying about their children participating in raves.

*Figure 10.2* Discoforum in 1980: the Swedish disco elite was handed awards from members of Village People

Source: Courtesy of Claes Hedberg

*Figure 10.3* Docklands raided by the Rave Commission

Source: Courtesy of Jan Quarfordt

## Curated by pioneers, spaces and resistance

For a product to find markets or to create a scene (such as club cultures), a variety of curatorial processes are involved. They set the framework by contextualising a specific phenomenon. Although recently applied in various other contexts (see Jansson and Hracs, 2018), the concept of curation originally stems from art and art markets (Balzer, 2014) where it has traditionally been associated with various processes from preserving and archiving to sorting, filtering, displaying and contextualizing artworks. Also, unlike traditional intermediaries, curators are recognised to be driven by motives other than pure economic (Joosse and Hracs, 2015; see also Comunian et al., this volume).

In the context of electronic dance music and the development of local and national scenes for club culture, this chapter has identified three actors or processes that in different ways filter, sort and contextualise (and thus set the limits for its expansion and impact) the music and the cultural expressions established in Stockholm during the studied period.

*Pioneers*: Ideas and products travel and they are diffused through individuals. The empirical examples show that specific pioneers were responsible for not only bringing influences from abroad, but also bringing physical artefacts like records. Crucially, these pioneers were primarily young and relatively resourceful people, with the means and/or knowledge of purchasing records, magazines and other media that conveyed familiarity with emerging electronic dance music. In contrast, the emerging rave culture in UK was largely driven by working class individuals, travelling by charter planes to Ibiza (Collins, 1998).

*Spaces*: The chapter has highlighted two dimensions of space. The first involves translocal flows. The second highlights the physical infrastructure and the importance of appropriate and low-cost spaces for establishing musical genres and cultural expressions. It is shown that both public infrastructure in the form of municipal and state-financed premises as well as various private initiatives constitute a space in which both disco and rave have been able to develop. Thus, space in itself may be an important curatorial dimension which acts in combination with people to serve as local anchors for social movements (Hracs and Jansson, 2020). The question of access to suitable premises for music related activities (such as concert venues, bars, discotheques) has been an ongoing debate, beginning with the introduction of electronically amplified music in the mid-1950s. For example, in 2019 an intensive debate regarding the noise level and outdoor activity that pubs, discos and concert venues generate is taking place and neighbours' complaints have on several occasions led to the closing of bars and venues in Stockholm (Andersson, 2019). Hence, although there has been an infrastructure for the introduction of electronic dance music, society and regulations have limited this development and continue to do so.

*Resistance*: On one hand, institutionalised resistance originating from mainstream society, policy and regulations may have hindered the introduction of electronic dance music. On the other hand, grassroots resistance against these regulations and attitudes has functioned as a cohesive force for these emerging subcultural movements. Indeed, these two forms of resistance may constitute an important contextualisation in such a way that they create identities consisting of a clearly defined 'us' and an easily identified opposition; a process of identification expressed by Buckland (2002) as 'the pleasure of opposition'. Similarly, media can act as a context and display of new movements and cultural expressions in both positive and negative ways. For example, being recognised in mainstream media "often has the effect of certifying transgression and legitimizing youth cultures", and at the same time, acceptance from the media and society may change the subcultural dynamics completely as "approving reports in mass media like tabloids or television, however, are the subcultural kiss of death" (Thornton, 1995, p. 6). Hence, in this context, resistance points to the subtle contestation between institutional forces, media and club cultures that filter, select and legitimise i.e. curate the development of subcultural expressions.

## Conclusion

In this chapter, an extensive range of empirical material has been analysed using the conceptual framework of curation to understand the development of electronic dance music in Stockholm. By demonstrating how pioneers, spaces and resistance are constituting a context that curates the supply and demand of subcultural products and expressions, this chapter contributes to a research agenda that nuances our knowledge of the temporal and spatial conditions in which cultural and creative industries are nourished and, more specifically, where electronic dance music cultures are developed. Applying the concept of translocality to spaces and flows has also enabled the chapter to engage with the tension between isolated and interconnected spaces of creativity. By extension, the chapter addressed the tension between individual and collaborative creative practices by showing how complex and shifting networks of actors underpin the production and consumption of cultural products.

## References

Andersson, O. (2019). En storstad måste vara till för alla. *Dagens Nyheter*, Aug. 18.

Balliger, R. (1995). Sounds of resistance. In: R. Sakolsky and F. Wei-Han Ho, eds., *Sounding Off! Music as subversion/resistance/revolution*. Brooklyn, NY: Autonomedia, pp. 13–26.

Balzer, D. (2014) *Curationism: How curating took over the art world and everything else*. Toronto: Couch House Books.

Brewster, B. and Broughton, F. (1999). *Last night a DJ saved my life*. London: Headline Book Publishing.

Buckland, F. (2002). *Impossible dance: Club culture and queer world-making*. Middletown, CT: Wesleyan University Press.

Caves, R. (2000). *Creative industries—Contracts between art and commerce*. Cambridge, MA: Harvard University Press.

Collins, M. (1998). *Altered state: The story of ecstasy culture and acid house*. London: Serpent's Tail.

Cresswell, T. (2006). You cannot shake that shimmie here: Producing mobility on the dance floor. *Cultural Geographies*, 13(1), pp. 55–77.

Crewe, L. and Beaverstock, J. (1998). Fashioning the city: Cultures of consumption in contemporary urban spaces. *Geoforum*, 29(3), pp. 287–308.

Currid, E. and Connolly, J. (2008). Patterns of knowledge: The geography of advanced services and the case of art and culture. *Annals of the Association of American Geographers*, 98(2), pp. 414–434.

Gavanas, A. (2020). *Från Diskofeber till Rejvhysteri i den svenska dansmusikhistorien*. Stockholm: Stockholmia.

Gavanas, A. and Öström, A. (2016). *DJ-liv. Historien om hur diskjockeyn erövrade Stockholm*. Möklinta: Gidlunds.

Gibson, C. (1999). Subversive sites: Rave culture, spatial politics and the internet in Sydney, Australia. *Area*, 31(1), pp. 19–33.

Grabher, G. (2002). The project ecology of advertising: Talents, tasks, and teams. *Regional Studies*, 36, pp. 245–262.

Greiner, C. and Sakdapolrak, P. (2013). Translocality: Concepts, applications and emerging research perspectives. *Geography Compass*, 7(5), pp. 373–384.

Hall, S. and Jefferson, T. (2006). *Resistance through rituals. Youth subcultures in post-war Britain*. London: Routledge.

Hesmondhalgh, D. (2002). *The cultural industries*. London: Sage.

Hodkinson, P. (2004). Translocal connections in the goth scene. In: A. Bennett and R. Peterson, eds., *Music scenes: Local, translocal and virtual*. Nashville: Vanderbilt University Press.

Hracs, B.J. and Jansson, J. (2020). Death by streaming or vinyl revival? Exploring the spatial dynamics and value-creating strategies of Stockholm's independent record shops. *Journal of Consumer Culture*, 20(4), pp. 478–497.

Hracs, B.J. and Webster, J. (2020). From selling songs to engineering experiences: Exploring the competitive strategies of music streaming platforms. *Journal of Cultural Economy*, 14(2), pp. 240–257.

Jansson, J. and Hracs, B.J. (2018). Conceptualizing curation in the age of abundance: The case of recorded music. *Environment and Planning A: Economy and Space*, 50(8), pp. 1602–1625.

Johansson, O. (2020). *Songs from Sweden: Shaping pop culture in a globalized music industry*. Singapore: Palgrave Pivot, Springer.

Joosse, S. and Hracs, B.J. (2015). Curating the quest for 'good food': The practices, spatial dynamics and influence of food-related curation in Sweden. *Geoforum*, 64, pp. 205–216.

Latour, B. (1987). *Science in action: How to follow scientists and engineers through society*. Cambridge, MA: Harvard University Press.

Lorenzen, M. (2018). The geography of the creative industries: Theoretical stocktaking and empirical illustration. In: G. Clark, M. Feldman, M. Gertler, and D. Wójcik, eds., *The new Oxford handbook of economic geography*. Oxford: Oxford University Press.

Maly, I. and Varis, P. (2015). The 21st-century hipster: On micro-populations in times of superdiversity. *European Journal of Cultural Studies*, pp. 1–17.

Molotch, H. and Treskon, M. (2009). Changing art: SoHo, Chelsea and the dynamic geography of galleries in New York City. *International journal of Urban and Regional Research*, 33(2), pp. 517–541.

Oakes, T. and Schein, L. (2006). Translocal China—an introduction. In: *Translocal China: Linkages, identities, and the reimaging of space*. London: Routledge.

Peterson, L. (1980). Sweden is invaded. USA is taking over with mass culture as its weapon. *Aftonbladet*, June 28.

Power, D. and Hallencreutz, D. (2007). Competitiveness, local production systems and global commodity chains in the music industry: Entering the US market. *Regional Studies*, 41(3), pp. 377–389.

Power, D. and Jansson, J. (2004). The emergence of a post-industrial music economy? Music and ICT synergies in Stockholm, Sweden. *Geoforum*, 35, pp. 425–439.

Power, D. and Scott, A., ed. (2004). *Cultural industries and the production of culture*. London: Routledge.

Thornton, S. (1995). *Club cultures: Music, media and subcultural capital*. Cambridge: Polity Press.

Wilderom, R. and van Venrooij, A. (2019). Intersecting fields: The influence of proximate field dynamics on the development of electronic/dance music in the US and UK. *Poetics*, 77, Dec., p. 101389.

# Index